The Intelligent Marketer's Guide to Data Privacy

Robert W. Palmatier · Kelly D. Martin

The Intelligent Marketer's Guide to Data Privacy

The Impact of Big Data on Customer Trust

Robert W. Palmatier
Foster School of Business
University of Washington
Seattle, WA, USA

Kelly D. Martin
Colorado State University
Fort Collins, CO, USA

ISBN 978-3-030-03723-9   ISBN 978-3-030-03724-6   (eBook)
https://doi.org/10.1007/978-3-030-03724-6

Library of Congress Control Number: 2018961178

Cover image: © Leon Harris/Cultura/Getty
Cover design by Tjaša Krivec

This Palgrave Macmillan imprint is published by the registered company Springer Nature Switzerland AG
The registered company address is: Gewerbestrasse 11, 6330 Cham, Switzerland

Preface

In 2013, the authors of this book had a casual conversation about the tension between "big data" and "privacy" and our sense that it might become a big deal for both marketing thought and practice. Inspired by these early chats, we embarked on a series of formal research projects, through which we have learned that for customers and marketers, big data and their implications on privacy are a really big deal. With this research, we have taken a programmatic approach to investigate deeply both the customer and firm outcomes of data privacy, gleaning insights that should be relevant for a wide academic audience. But we also did not want to limit the insights to academia, because the implications of data privacy for marketers who live these tensions daily seemed too important.

With this realization, we also conducted a thorough search for relevant, practical materials on data privacy that could help marketing managers. We were struck by the extent to which most books adopt either a legal or a technical data security perspective. No one seemed to be talking about the *customer* in a way that was designed to help marketers do their jobs better. We wrote this book to introduce the collective insights we gathered from our joint research program but also to include insights from many other researchers, so that the ultimate result would be a text that provides real, practicable help to marketing and business managers. In addition to our various data privacy collaborations, as an author team, we have wide-ranging expertise in customer engagement, relationships, and trust, as well as marketing's interface with society and ethics. Together then, we bring a unique perspective to data privacy that puts customer interests first, in a way that makes sense given the business realities of today, including the clear need to use marketing analytics and customer insights.

We structure this book from the vantage point of the marketing manager, putting ourselves in her or his shoes to explain the basic concepts necessary for a strong foundation in data privacy. The book begins with an overview of the data privacy tensions marketers grapple with, making the case for why this tension is so critical—and why it is not likely to go away. We also dispel some common myths about customer behavior and its inconsistencies, which has notable relevance for marketing practice. We spend an entire chapter within the mind of the customer and how privacy appears in that setting. Our goal is not to review the vast psychology research on this topic, but our summary offers a glimpse into how customers think about the data privacy practices that are most likely to matter to marketers. We take a similar approach to our legal overview in Chapter 3. A regulatory lawyer, Glory Francke, authored this chapter, based on her experiences practicing privacy law in the European Union and the United States. This concise overview covers the relevant laws and regulations that matter most to marketing managers and how they actually perform their jobs. The remainder of the book then presents both defensive and offensive strategies for managing data privacy. It was important to us to offer recommendations that are both practical and actionable, and that are backed up by concrete research findings and not merely a collection of anecdotes and stories.

As we explain in the opening chapter, we were driven to write this book because we know that marketers strive to do the right thing by their customers. The complex landscape of data privacy, its marketing applications, and its many customer implications sometimes make the path to doing the right thing obscure and difficult to find. We hope to clear the path, in a way that benefits the customer and, ultimately, creates better business practice. We will also continue posting new research and writings on this subject for use by our readers at SAMSinstitute.com.[1]

Seattle, USA	Robert W. Palmatier
Fort Collins, USA	Kelly D. Martin

---

[1]The SAMS Institute (Sales and Marketing Strategy Institute) is a voluntary association of professors and interested business executives who seek to *link academics to business for knowledge*.

Acknowledgements

We have benefitted from research collaborations with a variety of scholars who helped inform this book's central themes. We gratefully acknowledge Patrick Murphy, Abhishek Borah, Josh Beck, David Stewart, and O. C. Ferrell, whose important contributions have defined the academic literature on data privacy in marketing. We thank Glory Francke for her willingness to work with us and write Chapter 3, focused on privacy laws and regulations. Jisu Kim, Emma Giloth, Monika Flanagan, and Sydney Zeldes provided excellent research assistance. Both authors extend their appreciation to Elisabeth Nevins for editing and often rewriting the chapters to enhance the readability of this book. Financial assistance for our academic research came from the Monfort Family Foundation (Colorado State University) and the Marketing Science Institute (MSI). Robert Palmatier is also extremely grateful to Charles and Gwen Lillis for their support of the Foster Business School and his research, which helped make this book possible. Thank you for your generosity and support of our work.

Contents

About the Authors

Robert W. Palmatier is Professor of Marketing and John C. Narver Chair of Business Administration at the Foster School at the University of Washington. He founded and serves as the research director of the Sales and Marketing Strategy Institute (SAMSI), a global organization focused on linking business and academics to advance knowledge.

Robert earned his bachelor's and master's degrees in electrical engineering from Georgia Institute of Technology, as well as an M.B.A. from Georgia State University and a doctoral degree from the University of Missouri, followed by post-doctoral research at Northwestern University's Kellogg School of Management. Prior to entering academia, Professor Palmatier held various industry positions, including president and COO of C&K Components (global electronics company) and European general manager and sales and marketing manager at Tyco-Raychem Corporation. He also served as a U.S. Navy lieutenant onboard nuclear submarines.

Robert's research interests focus on marketing strategy, relationship marketing, customer loyalty, privacy, marketing channels, and sales management. His research has appeared in *Harvard Business Review, Journal of Marketing, Journal of Marketing Research, Marketing Science, Journal of Academy of Marketing Science, Journal of Retailing, Journal of Consumer Psychology, Marketing Letters,* and *International Journal of Research in Marketing.* He has also published several leading textbooks, including *Marketing Channel Strategy; Marketing Strategy: Based on First Principles and Data Analytics;* a monograph entitled *Relationship Marketing;* and chapters in numerous books. His research has been a highlight in *Nature, The New York Times Magazine, LA Times, Electrical Wholesaling, Agency Sales,* and *The*

Representor, as well as on NPR and MSNBC. In a recent AMA study, he was identified as the tenth most productive scholar in marketing over the past 10 years.

He has served as Editor-in-Chief of the *Journal of Academy of Marketing Science* and presently is the co-editor for *Journal of Marketing*; he also sits on numerous editorial review boards. His publications have received multiple awards, including the Harold H. Maynard (twice) and Sheth awards at *Journal of Marketing*, as well as the Robert D. Buzzell (twice) and Lou W. Stern (4 times) awards, selections as both an MSI Scholar and Young Scholar, the Varadarajan Award for Early Contribution to Marketing Strategy Research, and the American Marketing Association Best Services Article awards. As a teacher, he also has earned multiple awards for his courses on advanced marketing strategy in the doctoral, E.M.B.A., and M.B.A. programs at the University of Washington.

Among the numerous industry and governmental committees on which Robert has served, he chaired proposal selection committees for the National Research Council (NRC), National Academy of Sciences (NAS), and the Wright Centers of Innovation, which awarded grants of $20 million for the development of a new Wright Center of Innovation based on joint academic–industry proposals. He has served on NASA's Computing, Information, and Communications Advisory Group, with the AMES Research Center. This advisory group assesses the current state of technology development within academia, governmental agencies, and industry related to NASA's information technology activities and space exploration requirements. He also consults and serves as an expert witness for various companies, including Alston+Bird, Paul Hastings, Microsoft, Telstra, Starbucks, Emerson, Fifth Third Bank, Wells Fargo, Genie, Cincom, Tableau, Concur, World Vision, and Belkin.

Kelly D. Martin is Associate Professor of Marketing, Monfort Professor (2014–2016), and Dean's Distinguished Research Fellow at Colorado State University in Fort Collins. She holds a bachelor's degree from Gonzaga University, an M.B.A. from the Heider College of Business at Creighton University, and a Ph.D. from the Carson College of Business at Washington State University, where she was Outstanding Graduate Student for Research (2007).

Kelly's research interests involve marketing ethics and firm strategy, especially in the areas of data privacy, political marketing strategy, and consumer well-being. Her work has appeared in journals such as *Journal of Marketing*, *Academy of Management Journal*, *Journal of Consumer Research*, and *Harvard*

Business Review. These articles also have been recognized for their notable research impact, as recipients of the MSI Robert D. Buzzell Best Paper Award and Thomas C. Kinnear/*JPPM* Award, as well as finalists for *Journal of Marketing*'s Shelby D. Hunt/Harold H. Maynard Award, MSI/H. Paul Root Award, and Sheth Foundation Best Paper/*JAMS* Award. Kelly received the inaugural AMA Marketing and Society Emerging Scholar Award and held a Colorado State University Monfort Professorship (2014–2016), awarded for research promise. She earned the CSU College of Business Excellence in Research Award in 2013.

In addition to serving as the associate editor of *Business Ethics Quarterly* and *Journal of Public Policy & Marketing*, Kelly sits on the editorial boards of *Journal of Marketing* and *Journal of the Academy of Marketing Science.* She has been featured on American Public Media's *Marketplace Report* and Colorado Public Radio, and her work has been highlighted in *Nature, Strategy + Business, Science Daily*, and *Marketing News*, among others.

Kelly teaches Quantitative Business Analysis across Colorado State University's College of Business M.B.A. Programs, and she serves on the Oversight Committee for the M.B.A. in the Marketing Data Analytics Program. She has twice been awarded the CSU College of Business Excellence in (Graduate) Teaching Award and been nominated for the CSU Alumni Association Best Teacher Award and Beta Gamma Sigma Most Influential Teacher Award. Prior to academia, she worked as a marketing director in the assisted living industry.

List of Exhibits

Part I

Understanding Customer Data Privacy Fundamentals

1

Customer Data Privacy:
Why Every Marketer Should Care

Introduction

In 2014, the 113th U.S. Congress adopted a nonbinding agreement to
recognize January 28 as "National Data Privacy Day." Companies cel-
ebrate, or at least acknowledge, the annual observation by participat-
ing in various events, such as the speakers and panel discussions held at
LinkedIn's headquarters in San Francisco. Motivation for National Data
Privacy Day was nicely articulated by David Hoffman, Associate General
Counsel and Global Privacy Officer for the Intel Corporation who noted,
"All companies are data companies today, and we all have a role to play
in promoting the ethical and innovative use of data" [1]. This quote reinforces growing evidence that many prominent technology companies,
including Intel, welcome the event and embrace related data privacy initiatives. The Data Privacy Day Impact Report for 2018, published by the
National Cyber Security Alliance overviewed a variety of these initiatives
by its partner and supporting organizations. Another notable technology company mention included Facebook, which had recently launched
its Privacy Checkup tool and announced its commitment to simpli-
fying its privacy settings to give customers greater control over personal
information.

Just a short time after the 2018 Data Privacy Day though, we learned of
one of the most invasive consumer data privacy violations of our time, with
Facebook in the center of the maelstrom. An application developer and academic had worked to harvest highly sensitive personal information, using

© The Author(s) 2019
R. W. Palmatier and K. D. Martin, *The Intelligent Marketer's Guide to Data Privacy*,
https://doi.org/10.1007/978-3-030-03724-6_1

what was billed as a "personality quiz," which the highly sophisticated and savvy data analysis firm Cambridge Analytica purchased and used, allegedly for nefarious reasons—including exerting untoward influence over U.S. election outcomes. Facebook CEO Mark Zuckerberg endured days of Congressional testimony, answering myriad questions from various lawmakers about his company's failure to protect the personal information of its user base. Of gravest concern to Congress and the public at large was the realization that most of the information was being used without people's knowledge or consent. Only 270,000 Facebook users had agreed to install the "thisisyourdigitallife" app, but anyone linked to those users through networks of friends suffered compromised privacy, with contents of their personal messages readily available to the since-terminated Cambridge Analytica [2]. Ultimately, the breach is thought to have affected more than 87 million Facebook users.

Immediately after the scandal broke, many consumers immediately deleted the app or vowed to limit their personal use. Widespread coverage and detailed, step-by-step explanations circulated about how to delete Facebook from not only people's devices but also, more challengingly, from their lives. Several high profile companies and their CEOs—including the Firefox browser's parent company Mozilla, Tesla and SpaceX CEO Elon Musk, and Apple co-founder Steve Wozniak—noted that they personally had deleted their Facebook pages [3]. On the day of the breach announcement, the company's stock price dropped 7%, its largest drop ever, resulting in a loss of about $43 billion.

Furthermore, people familiar with the inner workings of Facebook's data harvesting capacity speculate that Cambridge Analytica may be the tip of the proverbial data misuse iceberg. Applications even more powerful than Cambridge Analytica's seems to be something of an open secret among marketers seeking sophisticated data techniques to target consumers and modify their behavior [4]. As the Cambridge Analytica scandal continues to unfold, growing public outrage about the blatant misuse and lack of customer transparency also has increased scrutiny into companies' data privacy practices. Experts have called for investigations of multiple companies' data privacy practices, noting that firms such as Google may have considerably more personal data than Facebook does [5]. For example, Google has ready access to people's search and browsing histories, installed apps, and shopping patterns, and by using identifiers, it can track users across all their devices.

And yet, even with all the bad press and critical coverage, within a few weeks, Facebook's financial health had fully recovered. Consumer ire over its mistakes continues to slowly dissipate. As one expert summarized the situation,

"The reality is that when it comes to privacy, the trade-off has already been made: We decided long ago to give away our personal information in exchange for free content and the ability to interact seamlessly with others" [6]. Even a known victim of the Cambridge Analytica data harvesting scandal, when asked if she would consider removing Facebook or constraining her use, admitted, "I'm just too nosy to stay off it" [2].

How Did We Get Here?

Increasingly, companies extract personal information and compile it in ways that their customers had not envisioned—regardless of what was stipulated in the company's privacy policy. Considerable ethical and legal questions arise, especially in relation to highly sensitive information (e.g., medical history, credit profile, sexual orientation, dating activity) that can be linked with great precision to people's general personal profile and then used in ways that people neither intend nor ever imagined. Consider a father who received a normally innocuous promotional mailer from Office Max. In what was likely a mail-merging mistake, his name was listed, followed by the identifying designation "daughter killed in car crash" [7]. Even companies with no apparent need for detailed personal information (unlike the data giants like Facebook and Google) are admitting to some issues for consumers, in the wake of seemingly endless reports of data breaches. Around the same time the Cambridge Analytica scandal broke, Panera Bread, Under Armour, Orbitz.com, and Saks Fifth Avenue all reported hacks that compromised millions of customer accounts [8]. Even if consumers are well-versed in cybersecurity, their data self-policing can offer only limited protection and privacy in the modern data-rich environment.

It also is important to remember that data privacy is not just an IT issue. In 2016, one marketing trade organization identified data security and privacy as the top "conundrum" for marketers [9]. Yet existing investigations tend to reinforce the belief that data privacy is not a marketing problem; many articles and books speak to just the technical or legal aspects of privacy. Business strategies designed explicitly to collect personal data, using marketing analytics (i.e., big data), and to monetize these customer data often fail to address the effects on customer relationships or other long-term marketing implications—to their detriment. We strongly believe that marketing should take a more focal position in this discussion, but we also recognize several explanations for the misdirected attention.

> Business strategies designed explicitly to collect personal data, using marketing analytics (i.e., big data), and to monetize these customer data often fail to address the effects on customer relationships or other long-term marketing implications—to their detriment.

Somewhat ironically, it may be that companies' current approaches to data collection and use reflect their own dedicated relationship formation efforts. As customers moved online and took a step away from brick-and-mortar transactions, it became harder for marketers to get to know their patrons without some other access to and sources of personal information. But good marketers must know their customers. In the early retail days, a shopkeeper could keep up with current events and familial happenings by chatting with customers [10]. These nostalgic encounters occur with far less frequency, but marketers' (and, to some extent, consumers') desire and need to form relationships is just as strong as ever. Data provide marketers with something of a way to "get to know" their customers. They can support precise targeting of products and services that are truly of interest to consumers, priced appropriately and communicated in places where customers are listening, ultimately delivered to people in a timely fashion and in their preferred mode.

Across industries, companies thus increasingly rely on customer data and marketing analytics to be effective. A survey commissioned by *Advertising Age* revealed that 90% of executives surveyed reported their dependence on customer data to conduct their marketing activities [11]. Companies take pride in offering personalized experiences to customers, and customers express appreciation for those offers by acknowledging the strong value they gain from personalized marketing. Public opinion polls signal that most consumers claim to guard their privacy closely, and are greatly concerned about online security, yet a survey by Accenture revealed that more than 60% of U.S. consumers also appreciate receiving relevant offers—and find that receipt even more important than keeping their information private [12]. In this sense, the initial motivation for harnessing insights from customers' personal information grew largely from the desire to forge strong, lasting relationships, connecting with consumers in valuable ways, to provide useful, free platforms for the benefit of the general public.

Another key cause of the current quandary, balancing total access to consumer data against reasonable checks and balances to protect personal information, reflects a totally different business mindset. As businesses became accustomed to receiving an unobstructed flow of customers' personal information, the view seemingly has shifted, such that consumers are no longer end-users or the objects of marketing attention but instead *consumers have become the product*. Specifically, consumers and their personal information provide what originally had been an unanticipated, or perhaps serendipitous, new profit source in markets populated by data brokers and third-party data retargeting firms that would pay a premium for the personal information that companies already possessed. The approximately 4000 data brokers in the United States today create seamless profiles of all of us, linking personal information that they have gathered from varied, disparate sources [5]. In this light, a darker, more sinister view of customer data privacy emerges. This view frames the consumer as powerless, subjected against his or her will to the whims of companies whose actions are covered, if dubiously, by increasingly complex and indecipherable privacy policies. To summarize, other than personal data, "There is no other asset class where the chief stakeholder has zero rights at the negotiating table" [13].

The reality likely falls somewhere between a view that suggests companies engage in altruistic efforts to connect with and serve customers, versus a more nefarious view in which companies treat customers as unwitting profit sources. Making clear sense of the psychological, legal, and practical tools available to marketers to manage data privacy in turn can be daunting. This obvious tension for marketers represents the inspiration for this book. Data privacy is an issue that transcends any single job title or discipline, and it is more critical than ever for business to get it right. We believe that marketers are well-positioned to make that happen, with unique capabilities, strengths, and backgrounds to create innovative data privacy solutions for their own companies and society at large. We have gleaned insights from our own research and practical experience with data privacy topics that speak directly to marketers, so with these lofty goals in mind, we seek to offer a comprehensive toolbox that will enable marketers to establish a more functional environment, in which consumer data privacy is a reality and a priority.

Why Data Privacy Matters

Research evidence strongly suggests that consumers desperately want better data privacy protections. When marketers simply collect customers' information—regardless of what they do with it—our research shows that people experience strong feelings of vulnerability [14]. Vulnerable customers are more likely to feel violated and lose trust in a company, ultimately leaving them prone to switch to a competitor. A recent poll shows data privacy is one of the few, rare topics on which most people agree. Even people with contrasting personal ideologies, such as Democrats and Republicans, agree that the current data privacy environment is inadequate. Citing Simmons Research, an NBC News feature revealed that 66% of Democrats and 68% of Republicans want more control over their personal information. Yet only 21 and 14% (respectively) trust the government to create solutions. As Chinni and Bronston ask,

> How often do you see members of the nation's two major political parties show that kind of agreement on anything, particularly a hot topic in the news? That's one reason why this issue isn't likely to go away…. Americans are nervous about sharing information online. They want more control over their data. And they aren't especially eager to scale back on the online time that has become a very big part of their lives. But they're not really sure they trust the government to figure out the answer. [15]

As we highlight in Chapter 2, polling averages hover around 90% when surveys ask people how much they worry about their data privacy, they sense a lack of transparency about how their personal information is used, or want to be able to manage the use of that information better. A recent poll indicates that 60% of people find cybersecurity issues more frightening than worries about going to war [16]. Even Facebook's Mark Zuckerberg, whose entire business model is built on getting users to relinquish their data privacy, admitted that more stringent regulation of data privacy practices was "inevitable" [17]. Yet to date, the U.S. Congress appears both unwilling and unable to issue any sort of data privacy regulations, which leaves individual companies' behavior even more critical.

The Privacy Paradox Is No Excuse

Consumers often get a bad reputation for inconsistency, due to the differences between their stated preferences and their actual behavior. They claim to value their data privacy greatly while simultaneously acting in ways that compromise that privacy, in what has been termed the *consumer privacy paradox*. People are chided for their foolishness in sharing considerable, highly personal information online, especially via social networks, and then acting surprised later when that personal information gets misused. In an argument critical of these naïve consumers, accusers assert that users of social platforms had no reasonable expectation of privacy in the first place. Furthermore, they claim

Consumers often get a bad reputation for inconsistency, due to the differences between their stated preferences and their actual behavior. They claim to value their data privacy greatly while simultaneously acting in ways that compromise that privacy, in what has been termed the *consumer privacy paradox*.

that people with "nothing to hide" have no reason to worry about widespread sharing and use of their personal information, so consumers simply have to accept the loss of data privacy [18]. Consumers can be their own worst enemies, if they accept or even embrace these ideas.

But a more likely explanation exists for the consumer privacy paradox, as is increasingly detailed in academic research [19]. That is, many people simply do not understand how to protect themselves and their personal information online where firms often make opting out difficult and maybe even purposefully so in some cases. In our research, we have found that even though consumers feel highly vulnerable when companies simply access their personal information, they also feel ill-equipped to manage effective data privacy protections by themselves. A Pew Research Cybersecurity Quiz echoes this assertion. The results from its short cybersecurity questionnaire, covering topics ranging from strong password identification to understanding email encryption, revealed that a substantial majority of the 1055 Americans polled could answer only 2 of the 13 questions [20]. As data privacy grows increasingly focal to the national conversation though, the consumer privacy paradox cannot remain a viable excuse that companies use to avoid responsibility for their insufficient consumer data privacy protections. Companies are being called on to do more.

It Is on Companies to Do More

Even before the Facebook scandal, even before the enforcement of more stringent privacy regulations across Europe, backlash to large-scale consumer data use had started growing. Business reporters implored the public to look beyond the most obvious purveyors of personal information and become more appropriately skeptical of an array of companies: "When you think about big data, you shouldn't just think about Google and Facebook; you should think about manufacturing and retail and logistics and healthcare" [21]. Equally forceful calls beg consumers to "take back" ownership and control of their personal information. When companies offer protection services, for a price, it becomes increasingly apparent to consumers what their personal information is worth. As they learn this value, or face requests to pay for its protection, consumers become better equipped to demand fair compensation for its use [22].

Exhibit 1.1: Walmart: A Company Doing More [23]

It is tempting to look only at bad behaviors and privacy failures. But it can be instructive to recognize how many companies are doing whatever it takes to get data privacy right. Take Walmart. In a recent webinar sponsored by the Marketing Science Institute (MSI), Walmart's Vice President and Chief Privacy Officer Jonathan Avila described how his company goes beyond mere principles of notice and consent to pursue a proactive, customer-focused approach that in turn establishes a strong data privacy culture.

Walmart operates under the general assumption that technological advances always outpace corresponding regulations. As we emphasize in Chapter 3, regulation is slow to emerge, and it often results only in response to negative customer outcomes or resounding advocate outcries. Noting these contrasting processes, Walmart proactively manages data privacy issues, with the understanding that to protect its valuable brand, it simply cannot afford a privacy failure. Damage to reputation is far more costly than any regulatory compliance would be.

With these ideas in mind, Walmart constantly monitors the technological landscape for potential threats to its customers and employees, raised by advocates or the public at large. Its dedicated privacy managers are well aware that technology will only become more complex, and their responsibility is to anticipate pitfalls. This forward-looking approach works so well to strengthen data privacy mainly because Walmart considers all technological changes and company practices according to a set of guiding values—one of which is respect for individuals.

Like other retail giants, Walmart also is successful because it can offer customers a seamless experience across online and offline worlds. Doing so demands extensive uses of data and analytics. Yet even this strategic imperative at Walmart is underpinned by its equally valued goal of being the most trusted retailer. Being trusted involves individual respect; at Walmart, that translates into respect for personal information.

To put these values into practice, Walmart embraces the principle of *privacy by design*. That is, customer data protections are not add-ons but rather must be built into the very essence of products and services, from their earliest idea phases. With a privacy-by-design mindset, concern for privacy infiltrates all of Walmart's organizational processes and structures. Every new product or service rolled out has been formulated with privacy in mind, leaving no room for unwelcome surprises.

We discuss these ideas in greater detail later in the book. This mindset can take any company beyond mere compliance to a culture that embodies privacy in everything it does. Companies that embrace privacy by design are subject to less regulatory concerns, so they ultimately benefit from greater flexibility. Finally, strong privacy can set companies apart from competitors, ensuring their lasting strategic advantage.

Most of the negative consequences for companies—such as having to devote resources to pay to obtain such personal information—are still to come. But several high-profile companies already have experienced significant

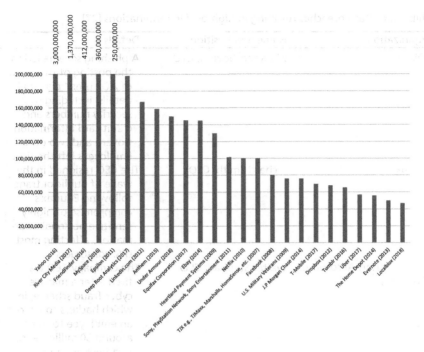

Exhibit 1.2 Top 25 company data breaches by number of records compromised

economic and public relations damage because of their failures to protect consumer information sufficiently, leading to extensive data breaches. In 2017, for example, 419 companies in 13 countries that had experienced significant data breaches (defined as more than 2600 compromised records) agreed to participate in IBM/Ponemon Institute's Data Breach Report [24]. The U.S. companies reported average losses of $7.35 million per breach. In our own research, we find that publicly traded companies' stock price drops, on the day of a breach announcement, by more than $8 million on average [25]. These losses can even spill over to close competitors that have not suffered a breach, so no modern company is immune from such events. Perhaps more frightening is historical evidence showing that nearly 30% of these firms will likely suffer another breach within 24 months of the initial occurrence [24]. Hitting particularly close to home for managers, in the wake of large-scale data breaches, companies face growing pressure to respond by announcing high-profile firings of one or more company leaders for their (perceived) failure to guard the company and its customers from the disasters [26] (Exhibits 1.2 and 1.3).

Exhibit 1.3 Data breaches resulting in high-profile terminations [27]

Organization	Name and Position	Details
AOL	Maureen Govern, CTO	A phishing scheme led to the posting of thousands of AOL users' personal details, including social security numbers and credit card information. Govern and two other employees left the firm
Equifax	Richard Smith, CEO	The CEO resigned in the wake of public outrage following Equifax's compromise of highly sensitive personal information, affecting most U.S. consumers
FACC, Austrian Aerospace Company	Walter Stephan, CEO	The company CEO was fired after a massive cyber fraud scheme, in which hackers convinced an employee to transfer around 50 million euro to a false account
HBGary Federal	Aaron Barr, CEO	Company and customer emails were hacked and shared. In addition to defacing the company website, hackers compromised the CEO's Twitter account and exposed his personal information
Home Depot	Frank Blake, CEO	Although Blake was planning to transition leadership around this time anyway, the timing of the breach created the appearance of yet another prominent CEO data breach firing
JPMorgan Chase	Jim Cummings, CSO Greg Rattray, CISO	After a massive data breach compromised more than 83 million accounts, the JPMorgan Chase CSO was reassigned, and its CISO was asked to step down

(continued)

Exhibit 1.3   (continued)

Organization	Name and Position	Details
KB Financial Group, NongHyup Card, and Lotte Card	Top executives	Top executives of each firm agreed to step down after a massive data breach compromising financial information
Maricopa County Community College District	Miguel Corzo, IT Director	Hackers breached the District's computer defense infrastructure, compromising the data of students, faculty, and other employees
Ohio University	William Sams, CIO	In addition to Sams, two top IT managers were fired after the university failed to prevent a series of breaches affecting students, faculty, and staff
Sony Entertainment	Amy Pascal, CEO	The company's embarrassing breach exposed the CEO's intimate personal communications about movies, actors, and so on. Some news sources say this firing was performance related, not due to the breach
Sony's Playstation Network	Howard Stringer	Following security breaches affecting gamers in the Playstation Network that exposed customer personal data, Stringer resigned
Target	Gregg Steinhafel, CEO Beth Jacob, CIO	Hackers breached Target's credit and debit account systems, accessing vast amounts of customer data during the 2013 Thanksgiving and Christmas shopping seasons
Texas State Comptroller's Office	Susan Combs, Comptroller	A data leak that exposed social security numbers, driver's license numbers, and other data led to an undisclosed number of executive officers being fired

(continued)

Exhibit 1.3   (continued)

Organization	Name and Position	Details
TJX	Gary Crittenden, Director Alexander Smith, Senior Executive VP	Both executives were forced to resign after the retailer's high profile breach, which affected millions of customers' credit and debit cards
Uber	Three top security managers	When Uber's hacking, and subsequent attempt to pay off the hackers, surfaced, the company fired three top security managers
Utah State Dept. of Technology	Stephen Fletcher, Department Head	Hackers breached the Utah Department of Health's servers to access thousands of Medicaid records in 2012
Yahoo	Ron Bell, Head Corporate Counsel	With the spread of details about Yahoo's data breach, the largest compromise of personal information ever, its head attorney stepped down. Yahoo failed to inform the public for years regarding this breach

Realm of Privacy and Data Collection

What do we mean when we discuss data privacy? The *Financial Times*, among others, equate *data privacy* with information privacy and note that it "concerns the collection, protection and dissemination of personal or private information about individuals or organizations." Yet there are problems with definitions that are so narrow in scope, and certainly all of us can think of situations or instances in which this definition would be inadequate. Such issues reflect the foundation for information privacy; the concept of *privacy* itself is notoriously difficult to define [17]. Recent commentaries [28] proclaim the elegant and timeless nature of the definition originally offered by legal scholars [29]. Privacy is simply the right to be let alone. Stewart advocates for this definition by emphasizing the equilibrium it creates, noting

"This is important for several reasons: (1) privacy is defined in terms of an individual's perception, being left alone; (2) the decision to share information rests with the individual; and (3) an invasion of privacy is a proactive act by another party" [26]. This definition also provides an appealing equilibrium, in that it refers to both who possesses the private information and how a potential violation or intrusion might occur.

Focus on Vulnerability

An imbalance in the personal data privacy equilibrium creates *vulnerability*, or feelings of susceptibility to harm. We argue throughout this book that minimizing consumer vulnerability is a much more realistic pursuit for marketers than is protecting privacy. Without a common agreement on what constitutes privacy, much less how to protect it, seeking to protect it is an unwinnable battle for companies that use customer data. A focus on minimizing customer vulnerability, however, is readily intuitive and actionable. It represents an area in which marketers can make the greatest positive impact. Our research, with thousands of customers and hundreds of companies over the years, has taught us that companies can minimize consumer vulnerability by being transparent in their use of personal data, as well as offering people some control over their personal data use. The combination is potent: When companies offer both *transparency* and *control*, customers feel empowered. Empowered customers trust the company more, feel less violated when information is shared, and report feelings of greater fairness, loyalty, and desire to share favorable word of mouth with others (Exhibit 1.4).

Our work has spanned industries, from financial to health care to technology to retail, and across the board, we consistently find that consumers crave both transparency and control. Organization size, age, structure, ownership, or location do not matter. The combined power of transparency and control for mitigating vulnerability transcends organizational characteristics or type. Our empirical analyses also show that transparency and control reduce the sense of vulnerability for people of all ages, education and income levels, geographic locations, and technological abilities. Yet a non-significant factor in our studies is measures of privacy concern. That is, mitigating privacy concerns does not help firms achieve beneficial outcomes; only reducing vulnerability does. This book therefore seeks to shift our thinking, from questions of privacy to concrete approaches for minimizing customer data vulnerability.

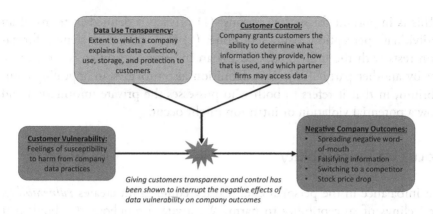

Exhibit 1.4   Transparency and control interrupt customer vulnerability

Plan for This Book

Based on what we have learned in our own research, as well as what we have gleaned from relevant academic findings and public opinion research, we offer a guide for smart marketers, who understand that doing something about data privacy is both important and no longer optional. This book is unabashedly geared toward marketers and business managers facing the reality of the modern "big data" revolution. We know marketing has passed the point of no return; actively using customer data to gain important insights is part of routine business operations. We also know that most marketers want to do the right thing by their customers. What marketers have not known thus far is how to walk this fine line. The structure of this book hinges on this very tension. We aim to provide marketers with the information they need to gain a solid, foundational understanding of data privacy concepts, along with strategies for both avoiding privacy failures and proactively using privacy as a strategic differentiator.

To accomplish these goals, this book is organized as follows: To break down the relevant consumer psychology behind data privacy, Chapter 2 contrasts how academic literature on consumer psychology fits with a trove of information from public opinion. We reconcile some of the contradictory findings by detailing how our own research on vulnerability, transparency, and control can illuminate how people feel about companies' uses of their personal information. A natural complement to this chapter follows in Chapter 3, where we explain cutting-edge privacy regulations. We give marketers the critical information they need to understand regarding legal compliance, here and abroad. We suggest ways intelligent marketers can exceed pure compliance and differentiate themselves from competitors, subject to the same rules and constraints.

The next section of the book, spanning Chapters 4–6, describes ways companies can guard against data privacy failures. Chapter 4 dives deeply into the extent to which customer relationship management (CRM) relies on customer data and accordingly outlines best practices for managing these critical systems. We suggest ways to understand the optimal balance between collecting customer data and minimizing vulnerability, while still optimizing the insights marketers gain so that they can connect with the right people. Chapter 5 expands on ways to guard against customer vulnerability. We explore the concepts of transparency and survey insights from various fields to provide marketers of all backgrounds with concrete ways to protect against damage to their customer relationships due to avoidable data privacy mistakes. Finally, Chapter 6 outlines options to help marketers recover after a data breach. No marketer wants to admit a data breach is possible, but the marketplace reality suggests they occur with some frequency. This chapter explains recovery strategies—what works and what has failed— to reconnect with customers, preserve important relationships, and restore reputation.

The final section offers ideas about competing by leveraging privacy. Chapter 7 includes actionable ideas for valuing customer data. We explore prior approaches to it, firm-side valuations, and consumer valuations of what they think their personal information is worth across different industry segments. In Chapter 8, we detail a data privacy audit that intelligent marketers can use to take stock of what their companies do to protect data privacy. This chapter and the audit tool provide an important application that allows marketers to gauge how their company's practices compare with others, including where they excel and where they might be deficient. The audit also establishes a useful framework for making sense of the remainder of the book. Because it exposes weaknesses and suggests opportunities for strategic differentiation, we explain in this chapter why it is so critically important for companies, of all sizes and across industries, to conduct regular data privacy audits that determine what data are needed, what they are collecting, and how those data are being stored and protected. The results can suggest where firms might need to cease collecting data that they never use or do not need. This approach is grounded in a basic principle: Data you do not collect cannot be breached or otherwise misused. The accessible data privacy audit in Chapter 8 can help marketers accomplish these goals.

Finally, in Chapter 9, we outline strategies that firms can use to leverage their data privacy practices and differentiate themselves from competitors. We regard the marketplace as primed and ready for organizations that truly understand and care about minimizing customers' vulnerability to data privacy breaches, as a way to create lasting competitive advantages. Better data privacy practices can have the unintended, paradoxical, but promising consequence of strengthening customer relationships—an outcome companies once might have sought by engaging in surreptitious uses of personal information.

We realize that marketers face a perpetual tension, between the desire to track everything about customers and extracting just the right amount of information to formulate the key insights needed to acquire and keep customers. Yet the reality is that people experience strong, negative emotional and cognitive responses immediately when companies access their data—whether or not they subsequently use those data. Customers crave simple, practical explanations of what companies are doing with their data, and they want a sense of control over the uses of those data. This book represents the first step to help smart marketers achieve greater customer empowerment, by shifting their thinking to imagine data privacy as a source of lasting competitive advantage.

References

1. National Cyber Security Alliance. (2018). *Data Privacy Day Impact Report 2018* [Ebook] (1st ed.). Retrieved from https://staysafeonline.org/wp-content/uploads/2018/03/2018-DPD-Report.pdf.
2. Frenkel, S., Matthew R., & Confessore, N. (2018, April 10). Facebook Data Collected by Quiz App Included Private Messages. *New York Times*. Available at https://www.nytimes.com/2018/04/10/technology/facebook-cambridge-analytica-private-messages.html?action=click&contentCollection=Technology&module=RelatedCoverage&region=Marginalia&pgtype=article. Accessed April 11, 2018.
3. Jenkins, A. (2018, March 28). These Companies Have Cut Their Ties with Facebook Amid the Cambridge Analytica Data Scandal. *Time*. Available at http://time.com/5216291/facebook-cambridge-analytica-companies-advertising/. Accessed April 10, 2018.
4. Hetherington, J. (2018, March 20). Facebook Whistle-Blower Reveals Cambridge Analytica Is Tip of Data-Mining Iceberg. *Newsweek*. Available at http://www.newsweek.com/cambridge-analytica-just-one-hundreds-mining-data-facebook-852744. Accessed April 10, 2018.

5. Mims, C. (2018, April 22). Who Has More of Your Personal Data Than Facebook? Try Google. *Wall Street Journal.* Available at https://www.wsj.com/articles/who-has-more-of-your-personal-data-than-facebook-try-google-1524398401?mod=trending_now_4. Accessed April 23, 2018.
6. Singer, N. (2018, April 11). What You Don't Know About How Facebook Uses Your Data. *New York Times.* Available at https://www.nytimes.com/2018/04/11/technology/facebook-privacy-hearings.html. Accessed April 11, 2018
7. Beckett, L. (2014, June 13). Everything We Know About What Data Brokers Know About You. *ProPublica.* Available at https://www.propublica.org/article/everything-we-know-about-what-data-brokers-know-about-you. Accessed April 24, 2018.
8. Paul, K. (2018, April 3). Everything You Wanted to Know About Data Breaches, Privacy Violations and Hacks. *MarketWatch.* Available at https://www.marketwatch.com/story/at-what-point-should-you-be-concerned-about-a-data-breach-2018-04-03. Accessed April 24, 2018.
9. Manion, J. (2016, January 20). Data Security and Privacy: Marketing's Top Conundrum in 2016. *MarTech Today.* Available at https://martechtoday.com/data-security-privacy-marketings-top-conundrum-2016-159239. Accessed April 17, 2018.
10. Martin, K. D., & Murphy, P. E. (2017, March). The Role of Data Privacy in Marketing. *Journal of the Academy of Marketing Science, 45*, 135–155.
11. Kaye, K. (2013, July 23). Study: Data Crucial to Marketing but Expensive to Safeguard. *AdAge.* Available at http://adage.com/article/datadriven-marketing/brands-consumer-data-safeguards-dma/243225/?utm_source=update&utm_medium=newsletter&utm_campaign=adage&ttl=1375196439. Accessed April 17, 2018.
12. *Digital Marketing Institute.* (2017). Why Marketers Should Care About Privacy Rulings. Available at https://digitalmarketinginstitute.com/blog/2017-8-03-why-marketers-should-care-about-privacy-rulings. Accessed April 17, 2018.
13. Davidson, J. (2018). *Forget Facebook, Meet the Company That Will Pay You for Your Personal Data.* Retrieved from http://time.com/money/3001361/datacoup-facebook-personal-data-privacy/.
14. Martin, K. D., Borah, A., & Palmatier, R. W. (2017, January). Data Privacy: Effects on Customer and Firm Performance. *Journal of Marketing, 81*, 36–58.
15. Chinni, D., & Bronston, S. (2018, April 15). Why Data Privacy Could Unite Red and Blue American. *NBC News.* Available at https://www.nbcnews.com/politics/first-read/why-data-privacy-could-unite-red-blue-america-n866081. Accessed April 16, 2018.
16. Rosenbush, S. (2018, April 16). The Morning Download: Companies Should Beware Public's Rising Anxiety Over Data. *Wall Street Journal.* Available at https://blogs.wsj.com/cio/2018/04/16/the-morning-download-companies-should-beware-publics-rising-anxiety-over-data/. Accessed April 19, 2018.

17. O'Brien, M., & Jalonick, M. C. (2018, April 11). Is Facebook Regulation 'Inevitable'? Not so Fast. *ABC News*. Available at https://abcnews.go.com/Technology/wireStory/facebook-regulation-inevitable-fast-54403103. Accessed April 19, 2018.

18. Solove, D. (2011). *Nothing to Hide: The False Tradeoff Between Privacy and Security*. New Haven, CT: Yale University Press.

19. Baruh, L., Secinti, E., & Cemalcilar, Z. (2017). Online Privacy Concerns and Privacy Management: A Meta-Analytical Review. *Journal of Communication, 67*, 26–53.

20. Olmstead, K., & Smith, A. (2017). What the Public Knows About Cybersecurity. *Pew Research Center*. www.pewinternet.org.

21. Tarnoff, B. (2018, March 14). Big Data for the People: It's Time to Take It Back from Our Tech Overlords. *The Guardian*. Available at https://www.theguardian.com/technology/2018/mar/14/tech-big-data-capitalism-give-wealth-back-to-people?CMP=share_btn_link. Accessed April 19, 2018.

22. Yokubaitis, R. (2018, January 19). It's Time to Take Our Privacy Back from Tech Companies. *The Hill*. Available at http://thehill.com/opinion/technology/369573-its-time-to-take-our-privacy-back-from-tech-companies. Accessed April 19, 2018.

23. Marketing Science Institute. (2018, January 19). Marketing at a Privacy Crossroads, Featuring Jonathan D. Avila, Vice President and Chief Privacy Officer, Stores, Inc. *Best of MSI Webinar Series*. Available at http://www.msi.org/video/marketing-at-a-privacy-crossroads/.

24. IBM Security, Ponemon Institute. (2017). *2017 Cost of Data Breach Study*. Available at https://www.ibm.com/security/data-breach. Accessed March 1, 2018.

25. Martin, K. D., Borah, A., & Palmatier, R. W. (2018, February 15). A Strong Privacy Policy Can Save Your Company Millions. *Harvard Business Review*. www.hbr.org.

26. Fuhrmans, V. (2017, October 12). New Worry for CEOs: A Career-Ending Cyberattack. *Wall Street Journal*. Available at https://www.wsj.com/articles/cybersecurity-tops-priority-list-for-ceos-after-string-of-high-profile-hacks-1507821018. Accessed April 19, 2018.

27. Table Compiled from Various Sources Including Armerding, T. (2018, January 26). The 17 Biggest of the 21st Century. CSO. Available at https://www.csoonline.com/article/2130877/data-breach/the-biggest-data-breaches-of-the-21st-century.html. Accessed April 20, 2018; Acuant, "40 Biggest Data Breaches of All Time". Available at https://www.acuantcorp.com/40-biggest-data-breaches-time/. Accessed April 20, 2018; and the Privacy Rights Clearinghouse at https://www.privacyrights.org/data-breaches.

28. Stewart, D. (2017, March). A Comment on Privacy. *Journal of the Academy of Marketing Science, 45*, 156–159.

29. Warren, S. D., & Brandeis, L. D. (1890). The Right to Privacy. *Harvard Law Review, 4*(5), 193–220.

2

The Psychology of Consumer Privacy

Introduction

Experts bemoan consumers' overall lack of concern about privacy and counterintuitive or even "paradoxical" behavior that exposes them to potential privacy violations. Vast evidence shows us that though consumers say that they are worried about privacy, they do not take the precautionary measures required to protect their personal information. These consumer-levied critiques become especially salient every time a large data breach or corporate privacy failure, such as the recent Facebook–Cambridge Analytica scandal or Equifax data breach, comes to light. Yet beyond these surface-level criticisms, far less attention has been devoted to the deep psychology of privacy. Bruce Schneier, the renowned author and security technologist, perhaps put it best when he wrote, "Privacy isn't dead, but it's certainly not healthy. And fighting for it can often feel futile. This is because we're fighting on the wrong playing field. We're debating the technologies while ignoring the psychology. And although privacy definitely is a technology problem, it's even more of a people problem" [1]. We agree, and hence, aim to fill this void by exploring this psychology. But while this chapter cannot offer a comprehensive psychological account of all things privacy, it does aim to provide marketers with a clearer understanding of customer mindsets, moving beyond the conventional wisdom or convenient stereotypes of consumers' seemingly ill-informed privacy behaviors.

© The Author(s) 2019
R. W. Palmatier and K. D. Martin, *The Intelligent Marketer's Guide to Data Privacy*,
https://doi.org/10.1007/978-3-030-03724-6_2

The lack of a clear perspective on consumer privacy might not be all that surprising, considering how many polls and surveys regularly make headlines, telling contrasting tales. In 1991, *Time*-CNN conducted a public opinion poll about privacy that showed 93% of consumers thought companies should be required to obtain their permission before selling their personal information [2]. Since then, this number has hovered around the 90% mark in responses to this and similar privacy questions, issued by a variety of polling and consumer research organizations. For example, a 1997 *Money Magazine* poll showed that 88% of the public favored a "Privacy Bill of Rights" that would require companies to specify what consumer information they collect and how they use it [3]. A *BusinessWeek*/Harris poll in 2000 revealed that 86% of Internet users want opt-in consent requirements for all personal information, including names, addresses, phone numbers, and financial information—all collected frequently by websites [4]. A 2015 University of Pennsylvania study indicated that 91% of Americans disagreed that company discounts are fair exchanges for personal information without consent [5]. According to *Consumer Reports*, 85% of respondents would refuse to provide even anonymous personal data in exchange for targeted advertising [6]. In Chapter 7, we report the results of our study on what consumers see as a fair price for their data across different situations (i.e., to improve their user experience, to receive more targeted advertising, and to sell to third parties).

As compiled by the Electronic Privacy Information Center (epic.org), AARP, Georgia Tech, Gallup, AT&T Research and others offer evidence that people overwhelmingly crave transparent privacy practices, but they sense little control over when and how their personal information gets used. Specifically, as part of the Pew Internet Research Program, researchers report that 9 in 10 U.S. adults believe that having control over personal information is very important [7], in terms of both who has the information and what information is obtained. Nor does this finding appear to vary by age or by gender. Only 9% of respondents to their study asserted that they had a lot of control over the information collected about them and its uses. Notably, social media users believe they have more control over their information than non-social media users. Finally, of the various agencies or groups that might collect and use customer data, people express the least amount of trust and confidence that online marketers can keep those data safe or use them in responsible ways. These findings are consistent with an expert's predictions about the future of privacy: "Public norms will continue to trend toward the desire for more privacy, while people's actions will tend toward giving up more and more control over their data" [8].

Yet we also find reason to believe that consumers lack the necessary information to make thoughtful decisions about their personal data. For example, a Gallup poll showed that 70% of U.S. adults believed the U.S. Constitution guaranteed their right to privacy; in fact, it offers no such provision [9]. A recent YouGov poll showed that a full one-quarter of U.S. respondents did not believe that they were at risk of hacking but also that their data or personal information were either very or somewhat vulnerable [10]. And then nearly 20% of the respondents to this study reported using no computer security protection at all, including the most basic measures such as using strong passwords and changing them frequently.

Queries about cybersecurity and data privacy also suggest a lack of consumer knowledge. In a Pew survey of 1055 U.S. adult Internet users, the 13-question quiz revealed vast inaccuracies in their descriptions of the best ways to keep their personal information safe [11]. Some performance differences arose, depending on their educational attainment, but across the board, people's knowledge was underwhelming. This lack of consumer understanding of data privacy also is coupled with a growing sense of pessimism and even helplessness. A parallel Pew study indicated that 9 of 10 adults believe they have lost control of how their personal information is collected and used [7]. In a 1998 AARP survey of members, 42% indicated that they did not even know to whom they would turn if a company misused or compromised their personal information [12].

With these contrasting realities in mind, this chapter begins with a brief overview of what social and environmental psychological theory has taught us about privacy over time. The theoretical evolution of privacy psychology in marketing has led to a largely rational framework for understanding consumer privacy, which we outline. But we also buttress this rational perspective and address its gaps, by emphasizing what we can learn from a more comprehensive account. This chapter closes with a nuanced perspective of consumer psychology in a time of unprecedented data privacy concerns.

Background: The Psychology of Privacy

The broad, general concept of privacy has been studied in social psychology and related disciplines for decades. One of the more famous definitions of privacy comes from Altman and refers to the ability to control access to one's self or relevant group (e.g., family) [13]. Environmental psychologists instead specify that privacy is an ability to control one's personal boundaries, and thus one's amount of contact with others [14]. Although this foundational work mainly was completed long before digital privacy concerns existed, a collective agreement establishes that privacy functions in society as an important human necessity, with critical benefits for individual persons, small-group dynamics, and society at large.

In a key paper, describing how privacy enhances societal processes and preserves individual well-being, Schwartz emphasizes the fundamental nature of privacy for protecting the human psyche [15]. He posits that just as people need to interact with one another and engage in various social circumstances, so too they require opportunities for withdrawal, which provides both "distraction and relief" from close peers or an uncomfortable level of social integration. A balance between social connectivity and privacy, he argues, preserves the healthy functioning of group interactions and situations. He thus goes on to mandate that "some provision for removing oneself from interaction and observation must be built into every establishment" (p. 741). Consumers' growing inability to extract themselves from the online environment, especially in a way that protects critical personal boundaries, suggests that it is more important than ever for marketers to understand the psychological effects of data privacy on consumers.

Schwartz also emphasizes that privacy guarantees are necessary for the stability and flourishing of any social system. As a strong advocate for a commonly held set of rules about privacy, whether codified or tacit, he suggests such norms can create a societal bond that facilitates more productive, healthy human interactions. Specifically, "rules as to who may and who may not observe or reveal information about whom must be established in any stable social system" (p. 742) [15]. Yet privacy involves not only self-protection and preservation benefits that enable social functioning, but it also is necessary for psychological well-being. Research into the psychological dimensions of privacy [14] reveals that privacy serves as an important human defense mechanism that creates a sense of balance and harmony. When achieved, these states allow people to develop and grow, better realizing their human potential. In its positive psychology, or developmental, sense, some minimal level of privacy is necessary for self-reflection and personal rejuvenation [16]. The benefits in turn include creativity, problem solving, and an ability to determine important life course directions.

Research into the psychological dimensions of privacy reveals that privacy serves as an important human defense mechanism that creates a sense of balance and harmony. [14]

This research stream also indicates that people perceive their privacy as most frequently violated in digital environments, where the negative consequences get intensified by perceptions of low control over who is managing their personal information or how, due to the essential structural features of the environment [16]. This problem is likely to be magnified as the *Internet of Things* (IoT) expands, with systems and digital personal devices that monitor and survey people's behavior continuously, providing no respite from some form of information collection and data extraction. As the digital environment gets further integrated into various aspects of people's lives, many find it "harder and harder to find a sphere of life where data are not collected, indexed, distributed, searched and inferred. The means for control, if provided at all, are opaque or incomprehensible" [17] (Exhibit 2.1).

Exhibit 2.1: The Helsinki Privacy Experiment [17]

Conversations about data privacy frequently segue to discussion about *surveillance*, which Schwartz defines as "institutionalized intrusions into privacy" (p. 742) [15]. With the proliferation of IoT technologies and more sophisticated tracking tools online, surveillance has become a real consumer fear. *Ubiquitous surveillance*, or the unilateral collection of personal data through sensors embedded throughout one's daily routines and environment [17], sparks significant consumer fear—but also is increasingly a reality, enabled by IoT technologies.

In a now famous study referred to as the Helsinki Experiment, researchers equipped 10 volunteers' households with video, audio, device tracking, smart sensors, and other technologies to create ubiquitous surveillance situations. The participants reported regularly on their stress levels and comfort with the surveillance over a six-month period. Although they initially expressed strong negative reactions to the technologies and the sense of constant monitoring, their feelings quickly subsided to just minor annoyance. After two weeks, participants reported that both their annoyance and anxiety had declined to the point that they had "gotten used to" the technology, and the surveillance began to feel normal.

Seven of the participants reported active privacy-seeking behaviors, such as using Internet cafes to avoid having their keystrokes tracked. Others changed their in-home behaviors, such as avoiding nudity, intimacy in personal relationships, or substance abuse in front of the cameras. But three respondents undertook no privacy-seeking behaviors whatsoever. For them, ubiquitous surveillance simply had become a normal and expected way of life.

A Rational Model of Consumer Privacy Psychology

Many scholars believe that a true understanding of privacy requires consideration of the context [18–21]. As this view has gained acceptance, either implicitly or explicitly, research has focused on studies that examine privacy in the context of their specific domain. For example, in marketing, we mostly investigate the psychology of privacy through the lens of people's roles as consumers. Marketing psychological perspectives thus refer to people analyzing, evaluating, purchasing, or otherwise interacting with goods and services. We review some of the more commonly used social psychological approaches, without claiming an exhaustive list. We organize these findings according to privacy concerns and their antecedents, privacy outcomes, and enhancing forces, as depicted in the rational model of consumer privacy psychology. This model reveals how consumer perceptions might shift, in relation to their determinants or consequences (Exhibit 2.2).

Considerations of *privacy concerns* in marketing research tend to reflect the widely held view that they offer a good proxy for how people feel about privacy, or their ability to control and thoughtfully decide what and how to share personal information [22]. This rational, sequential approach to studying consumer privacy has been so widely adopted that it earned the moniker *APCO:* antecedents, privacy concerns, and outcomes [19]. Yet subsequently research also has identified gaps [23] and sought to broaden the scope of this APCO model [24].

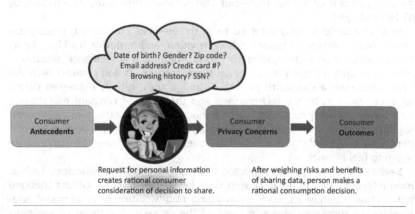

*Many privacy scholars, including the authors of this book, challenge the assumptions of the rational model of consumer privacy decision making.

Exhibit 2.2 Rational model of consumer Antecedents, Privacy Concerns, and Outcomes (APCO)

Consumer Heterogeneity

The potential antecedents, or factors that shape people's privacy concerns, are innumerable. Perhaps the vast number of personal characteristics, traits, and personality elements that can shape how people view privacy explain why privacy concern antecedents have been studied far less often than privacy outcomes. Reflecting this lopsided attention, our review of the antecedents necessarily is limited in nature; it also leads us to highlight that making sense of these various drivers remains an area ripe for additional investigation. In particular, as we will show, systematic understanding of the extent to which personal characteristics and traits do or do not affect privacy is far less well developed than other portions of this framework. Any interpretation of evidence about individual differences, antecedents, and other contingencies therefore requires some caution.

What we do know is that people's experiences with privacy events or abuses heighten their privacy concerns [19]. These experiences can shape their expectations for interacting with new technologies, which in turn influence their ultimate acceptance of those technologies [25]. As more consumers report being victims of data breaches or privacy violations, the influential privacy experience may have significant consequences for marketers. On the other side, personal agency and self-protection aspects, combined with an overall sense of personal control over information, can be helpful in reducing privacy concerns [26]. Communications from a company might increase or decrease consumers' privacy concerns, in conjunction with the various elements of company privacy policies [27–29]. Ultimately, privacy policies and communications about them can invoke feelings of reactance, justice, fairness, or (un)willingness to disclose personal information.

Personal psychological characteristics also may influence perceptions of the concept of privacy itself. Personality traits, often described according to Goldberg's Big Five personality factors (extroversion, agreeableness, emotional instability, conscientiousness, and intellect), determine how people process information and behave in response to it [30], including privacy-related information. According to extant research though, only agreeableness and emotional instability consistently and positively relate to privacy concern. These positive relationships tend to be stronger in health-oriented privacy contexts. In the workplace, employees who score higher on Machiavellianism scales (suggesting broadly that they believe the ends justify the means) and also work in R&D report greater acceptance of an employer violating their privacy. In contrast, idealistic employees consider privacy violations in the workplace unacceptable [31].

The patterns by which demographics shape privacy concerns appear far less predictable and even more dependent on the context and technology. This variability can be frustrating to marketers, for whom access to demographic profiles is often the easiest to attain, relative to more detailed personality or experience profiles. Perhaps the most widely misunderstood and misattributed consumer demographic is age. Conventional wisdom, supported by select media reporting, implies sweeping generalizations about young people's extreme willingness to sacrifice privacy for access to goods and services or their constant drive for greater connectivity with social communities. According to these overblown reports, young people share without any filters, across a myriad of social platforms, eschewing greater security protections even when they are readily available. One expert even went as far as to surmise,

> Young people are more used to a world with cameras everywhere. They spend more time online and identified. The older generation developed behavioral habits that assumed a degree of privacy that young people have not experienced. What oldsters would have to give up, young people will not miss. [8]

Admittedly, age can be influential in shaping privacy concerns in certain conditions and with regard to certain technologies. Yet in just as many settings, it is not significant. When age exerts an influence, it also might move in directions opposite what conventional wisdom predicts about older and younger consumers.

To obtain a complete picture of age, an ongoing challenge relates to the respondents who tend to join academic privacy research: They skew young. According to a recent meta-analysis, in studies that report the ages of their respondents, the median age is 24 years, and the modal age group ranges from 18 to 24 years if studies provide age ranges [32]. This meta-analysis includes 166 studies from 34 countries, with an overall sample size of more than 75,000 people. If these studies included more age variety, they might offer more in-depth insights; for example, privacy concerns seem to be increasing over time among consumers of various age groups, though perhaps more sharply for older consumers [33]. A couple of recent Pew Research studies affirm this trend. In one, six data surveillance and privacy scenarios posed to consumers evoked no significant differences across respondents of various ages [34]. In another, the researchers find that older adults continue to adopt new technologies despite their privacy worries, with rapid adoption growth even among those 65 years and older [35].

Another notable demographic factor is education; it appears to play a role in some perspectives on privacy but not others. For example, people with less education are significantly less aware and informed about how to keep themselves safe online, spanning the various cybersecurity topics on which they were queried [11]. Reviews of academic literature reveal that consumers who are less educated and poor tend to be less concerned about privacy [19]. Finally, women tend to express greater privacy concerns and restrain their technology use due to these concerns [32].

Consumer Privacy Concerns

In research that seeks to understand and measure consumers' feelings about information privacy and the related psychological constructs, privacy concerns have surfaced as an effective proxy. This work has gained more acceptance in fields such as information systems, however, than in marketing. Instead, early privacy research in the marketing domain relied on simply asking people about privacy and their privacy worries across different consumption contexts [36].

Research in marketing has shown that consumers' concerns about privacy can influence outcomes of relevance to marketers. They report decreased willingness to disclose information to companies and even signal reduced desire to purchase from previously preferred companies and brands. However, research has failed to produce a clear pattern of conditions under which consumers respond to privacy concerns in a predictable way. Information context, company privacy behavior, and the bond the customer has with the firm all can influence privacy concerns in ways that make research highly contingent.

Consumer Privacy Outcomes

To understand the effects of people's privacy concerns, marketers and consumer researchers frequently examine when customers are (are not) willing to share their personal information. The field commonly draws from useful frameworks such as social exchange and behavioral decision theories. But as we argue, these perspectives adopt a hyperrational view of customers that appears unrealistic against actual, often irrational consumer evaluation and behavior. Indeed, many privacy theories in the APCO family emphasize purposeful cost-benefit tradeoffs that consumers make prior to each privacy-relevant decision. This research often hinges on predictions that consumers weigh the consequences of their personal information disclosure against the value offered by the marketer. While some customers find the benefits outweigh costs and rationally decide to share their personal information, other customers derive a calculus that makes them unwilling to exchange personal information for greater convenience or other benefits.

Beyond willingness to share information, additional privacy outcomes investigated include consumer choice to provide a company with false information, clearly damaging to a firm relying on customer data insights or that monetizes such information. These behaviors can pose a substantial threat to firms that rely on accurate consumer information. As such, academic research often relies on reactance theory to explain consumers' responses to micro-targeting or other highly personalized messages that appear to violate their privacy [37]. When consumers receive overly targeted or uncomfortably personal marketing communications, they may avoid any further communication, provide false information, spread negative word of mouth, and so on [38]. Purchase intentions are another outcome of interest to both researchers and marketing managers. Large-scale field studies have shown that a company with stronger privacy policy enjoys increased purchase behavior [39].

The Important Role of Trust

As our book subtitle implies, trust plays a critical role in the customer-data privacy world. Trust has long been recognized as an important driver of positive marketing effects, including those reviewed above. We already know a lot about the role of trust in marketing, so it should come as no surprise that trust also benefits companies in data privacy contexts. It can function as an antecedent of privacy concerns, as well as a missing link between privacy concerns and people's willingness to engage with the firm on online and mobile platforms [40, 41]. Lack of trust in turn has negative effects on privacy-related issues. For example, organizations' covert use of various tracking and other personally-identifying technologies have been shown to damage trust [42]. Promoting trust, rather than trying to reduce privacy concerns, thus may represent a better use of firm resources [43]. With strong trust, firms attain beneficial outcomes, such as information disclosure and even stronger customer–firm relational bonds, even in industries or contexts that can make people feel vulnerable. Trust helps people appreciate highly personalized or targeted advertising content; lack of trust can cause highly individualized communications to seem creepy or disturbing [44]. Because creating customer trust has such myriad benefits for firms, notably in the realm of data privacy, we continue to focus on this aspect as we extend our discussion.

Beyond the Rational Model: A Better Way for Marketers to Think About Consumer Privacy

The rational psychological model has many strengths, but some obvious problems also arise if we assume solely rational consumer decision making when it comes to data privacy. It is easy to imagine situations in which people simply cannot make rational decisions about their personal information, such when their health is at risk or they must provide information for financial viability. When people suffer intense emotional states, their cognitive processing capacity diminishes [23]. Cognitive strain, which arises when people are cognitively taxed or tired, forces them to rely on less effortful options to make disclosure decisions, such that they might just give up and share personal information if doing so seems likely to help them reach a decision. During pleasurable interactions that provoke happiness, the high levels of oxytocin produced in the brain can lead people to disclose more information and ultimately put themselves at risk [45].

A better conceptual approach to understanding and addressing consumers' needs, assuming a *lack* of rational information processing, might be to conceive of the focal construct not as privacy but as vulnerability. In our research with consumers, we have found that people are not very good at understanding privacy in general, or data privacy in particular—regardless of the context [46]. What people are adept at identifying is when they feel vulnerable. We therefore propose consumer vulnerability as a superior approach to thinking about consumer privacy, for several reasons: It is readily definable and allows for intuitive categorization across various data privacy realms. Furthermore, vulnerability fits squarely in the gossip theory domain, which establishes a more solid theoretical foundation to make sense of how consumers really feel about their personal information being collected, analyzed, and shared.

> In our research with consumers, we have found that people are not very good at understanding privacy in general, or data privacy in particular—regardless of the context [46]. What people are adept at identifying is when they feel vulnerable. We therefore propose consumer vulnerability as a superior approach to thinking about consumer privacy.

Consumer Data Vulnerability

Vulnerability implies susceptibility to injury or harm [47]. When companies collect, store, and use consumers' personal information, it increases the potential for harm and thus their feelings of vulnerability. Marketers in turn must remember that *most negative consumer effects resulting from data use derive from people's anxiety about the potential for damage or feelings of violation, rather than actual data misuses or financial or reputation harm.* In our research, people report experiencing harm just from data access—regardless of whether their data subsequently have been analyzed, shared, or otherwise misused. Thus we consider it critical to understand the effects of consumers' feelings of vulnerability, rather than trying to determine how they understand the privacy elements associated with data use. Similarly, it is not enough for marketers to focus solely on actual or incurred consumer damages (Exhibit 2.3).

Exhibit 2.3: Going to Extremes: One Woman's Data Access Story [48]

In 2014, Princeton technology sociologist Janet Vertesi wrote for *Time* about her personal experience trying to hide her pregnancy from online retailers. Dr. Vertesi was studying ways people could retain possession of their personal information and decided to use her pregnancy as a personal experiment. She wanted to see if the conventional wisdom of the time, which prompted opt-out as an effective way to avoid being targeted, actually worked. What she found was that hiding an important life event (i.e., pregnancy information reportedly is worth more than the personal information of 200 other people) was far more difficult and complex than simply clicking an opt-out link.

Vertesi alerted her family and friends via phone that they were absolutely not allowed to mention her pregnancy online—especially on social media. A few violations nevertheless occurred, by relatives who truly did not understand that parts of Facebook were not private. They assumed that certain messaging approaches or ways of posting were protected, though they were not.

Then Vertesi downloaded the private browsing tool Tor, which consumers can use to avoid monitoring of their Internet searches, to search for baby names, maternity clothes, and items for the nursery. She bought all the baby items with cash or used prepaid Amazon cards and had her online purchases delivered to a locker. But her repeated purchases of gift cards prompted a warning that her behavior could be deemed suspicious and reported to the authorities. That is, her activities to keep her personal information secret from marketers had made her seem like a criminal!

Furthermore, opting out imposed higher prices and terrible inconvenience on this consumer. She thus closes her article with the following warning: "The data-driven path we are currently on—paved with the heartwarming rhetoric of openness, sharing and connectivity—actually undermines civic values and circumvents checks and balances. When it comes to our personal data, we need better choices than either 'leave if you don't like it' or no choice at all."

Consumer data vulnerability takes different forms, along a continuum of potential harm. The most benign form exists when companies just have access to a customer's personal data, or data access vulnerability. This mere access means that firms have "detailed digital dossiers about people" (p. 2) and can engage in the "widespread transfer of information between a variety of entities" (p. 2) [49]. Consumers limit how and with whom they share sensitive information to reduce this vulnerability, using disclosure management processes such as reactance or refusal. Yet companies already possess and continuously seek to increase customer information, so data access vulnerability remains a widespread and growing concern for consumers.

Data breach vulnerability increases consumers' perceptions of susceptibility to harm even more, because it implies a firm that already has their private data, or one of its close rivals, has suffered an actual security lapse. The U.S. Identity Theft Resource Center estimates that more than 1 billion personal records have been subjected to risk from data breaches (tracked from 2005 to date) [50]. Ultimately, not everyone whose records have been compromised experiences victimization, but the unknown scope and lack of control over this threat makes this type of vulnerability especially troubling to consumers.

Perceptions of vulnerability also increase following data breaches at firms that actually possess the customer's data (focal firm) but also, indirectly, after breaches at close competitors (rival firms), because these events increase the salience of the belief that such breaches are possible in that competitive domain. This spillover effect (or *spillover vulnerability*) arises when consumers perceive greater susceptibility to harm because a firm similar to one that has their data suffers a data breach. Spillover creates less vulnerability than a data breach at a focal firm, because even though their vulnerability becomes more salient, consumers perceive less vulnerability when a competitor suffers a breach, rather than a focal firm. Yet analyses of the effects of focal firm breaches clearly indicate damage to their close rivals. On the day of Nvidia's data breach in 2012, its rival Advanced Micro Devices (AMD) lost about $48 million, seemingly because of the similarities in the fundamental practices of AMD to Nvidia, suggesting their similar risk [51].

Finally, *data manifest vulnerability* occurs when consumer data actually are misused, which causes personal harm. Disclosures and fraudulent activities represent the most severe form of vulnerability, moving beyond susceptibility to a state of actual damage. Even if the actual damage a consumer experiences is minor, it significantly increases her or his perceptions of data vulnerability. Thus, the greatest effects tend to stem not from actual data misuse but from accompanying feelings of violation and the indeterminate nature of the threat (Exhibit 2.4).

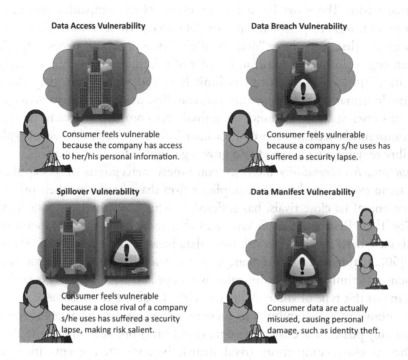

Exhibit 2.4 Data vulnerability types and definitions

Gossip Theory Insights

Consumers' psychological and behavioral responses to feelings of vulnerability can be informed by gossip theory, which refers to common, unsanctioned transmissions of personal information about a vulnerable third party. That is, *gossip* is any evaluative communication about an absent third party [52, 53], and gossip researchers report that approximately two-thirds of all communications in public social settings are devoted to such topics [54]. People are adept at detecting gossip, guarding against becoming a target, and minimizing their vulnerability to it [55, 56]. When they learn they are the target of gossip, people typically react negatively [57], with a range of negative emotional and cognitive responses [58], including heightened feelings of betrayal and violation [59] and deteriorating levels of trust [60]. An application of gossip theory to a business context suggests that consumer data vulnerability may lead to feelings of emotional violation and diminished cognitive evaluations of trust.

Gossip theory also identifies two factors that can suppress the negative effects of unsanctioned transmissions of information: transparency and control. *Transparency* allows the target to be aware of precisely which information is being shared and thus determine the scope of potential harm, which enables this target to develop strategies to counter any negative outcomes. *Control* instead is the extent to which the target believes she or he can manage the flow of information [54]. A perceived lack of control, evoked by learning about the transmission of personal information, exacerbates negative affect surrounding a gossip event, even if the valence of the information being spread is not negative [52]. As forms of empowerment, control and transparency thus may help people manage the negative effects of their own vulnerability [61].

Data Vulnerability Suppressors

Considering transparency and control in data privacy contexts can help marketers understand how to lessen the detrimental effects of consumer vulnerability. That is, in a consumer data vulnerability context, firms' data use *transparency*, or their efforts to explain their data collection, use, storage, and protection, and the extent to which they enable customer *control*, by granting consumers a means to determine what information they provide, how it gets used, and which partner firms may access that data, both are critical for mitigating the damaging effects. Both transparency and control exert beneficial effects on firm stock prices in the wake of a data breach; they also minimize people's feelings of vulnerability and violation and promote trust in large-scale consumer studies [46].

According to gossip theory, before a target can address a gossip event, he or she needs the knowledge that the gossip is occurring [62]. Thus, transparency should be a critical suppressor, with the potential to mitigate the harm wrought by consumer vulnerability, because it gives people the knowledge they need to evaluate the potential harm, including the nature and scope of the data that the firm possesses and how those data are used. Typically, firms provide transparency by issuing a privacy policy or information collection disclosure. In addition, company information collection strategies that are overt versus covert influence people's responses to their personalization efforts [44]. In a similar sense, transparency may be critical for firms to avoid the "creepiness factor" often associated with big data and analytical inferences about consumers [63].

Both transparency and control exert beneficial effects on firm stock prices in the wake of a data breach; they also minimize people's feelings of vulnerability and violation and promote trust in large-scale consumer studies. [46]

Providing control offers another key strategy. For example, when they learn that gossip has occurred, targets often seek to regain control of their information [54]. Salvaging this control also represents a key restorative element after a damaging gossip event [64]. Providing consumers with control enables them to manage and adjust their personal data preferences. To bestow control, firms might rely on opt-in and opt-out decisions [65] and give consumers tools to manage their individual settings and preferences, governing the use of their data. After Facebook suffered a data breach in 2010, it responded with new policies and systems that promised to "keep people in control of their information" [66]. As Facebook faces fresh scrutiny today, it continues to offer large-scale communications, suggesting to customers that it provides both transparency and control, to help mitigate damage and avoid their defection [67].

Beyond the unique suppressive effects of transparency and control, the most potent force for reducing the damaging effects of vulnerability on companies and their performance likely entails the combined effect of transparency *and* control. When targets seek to mitigate unsanctioned transmissions of their personal information, they rely on transparency and control concurrently. Strong transparency and control give people more knowledge of the firm's data management practices and the ability to manage their data portfolio. In contrast, consumers who only achieve transparency know of the potential harm but have no way to manage it; consumers who only have control can manage their data but have insufficient knowledge to make informed decisions.

References

1. Schneier, B. (2015). Fear and Convenience. In M. Rotenberg, J. Horwitz, & J. Scott (Eds.), *Privacy in the Modern Age: A Search for Solutions* (pp. 200–203). New York: The New Press.
2. *Time-CNN*. (1991). 93% of Respondents Believed That the Law Should Require Companies to Obtain Permission from Consumers Before Selling Their Personal Information. *EPIC.org. Public Opinion Polls Archive*. Available at https://www.epic.org/privacy/survey/#polls. Accessed March 31, 2018.

3. *Money Magazine.* (1997). Poll Shows That 88% of the Public Favors a Privacy Bill of Rights. *EPIC.org. Public Opinion Polls Archive.* Available at https://www.epic.org/privacy/survey/#polls. Accessed March 31, 2018.
4. BusinessWeek/Harris. (2000, March). A Growing Threat. *BusinessWeek Magazine.* Available at https://www.epic.org/privacy/survey/#polls. Accessed March 31, 2018.
5. Turow, J., Hennessy, M., & Draper, N. (2015). *The Tradeoff Fallacy: How Marketers Are Misrepresenting American Consumers and Opening Them Up to Exploitation.* Report from the Annenberg School for Communication, University of Pennsylvania. Available at https://www.asc.upenn.edu/sites/default/files/TradeoffFallacy_0.pdf. Accessed May 14, 2018.
6. Fox, J. (2014, May 27). *85% of Online Consumer Oppose Internet Ad Tracking, Consumer Reports Finds* (Consumer Reports). Available at https://www.consumerreports.org/cro/news/2014/05/most-consumers-oppose-internet-ad-tracking/index.htm. Accessed April 5, 2018.
7. Madden, M., & Rainie, L. (2015). Americans' Attitudes About Privacy, Security and Surveillance. *Pew Research Center.* www.pewinternet.org.
8. Rainie, L., & Anderson, J. (2014). The Future of Privacy. *Pew Research Center.* www.pewresearch.org.
9. *Gallup.* (1999). 70% of Respondents Believe the Constitution Guarantees Citizens the Right to Privacy. *EPIC.org. Public Opinion Polls Archive.* Available at https://www.epic.org/privacy/survey/#polls. Accessed March 31, 2018.
10. *YouGov.* (2018, February 10–12). Majority of Americans Feel Their Personal Information Online Is at Risk. Available at https://today.yougov.com/news/2018/03/12/majority-americans-feel-their-personal-information/. Accessed March 26, 2018.
11. Olmstead, K., & Smith, A. (2017). What the Public Knows About Cybersecurity. *Pew Research Center.* www.pewinternet.org.
12. *AARP.* (1998). Survey of Member Attitudes on Privacy. *EPIC.org. Public Opinion Polls Archive.* Available at https://www.epic.org/privacy/survey/#polls. Accessed March 31, 2018.
13. Altman, I. (1975). *The Environment and Social Behavior: Privacy, Personal Space, Territory, and Crowding.* Monterey, CA: Brooks/Cole Publishing.
14. Pedersen, D. M. (1997, July). Psychological Functions of Privacy. *Journal of Environmental Psychology, 17,* 147–156.
15. Schwartz, B. (1968, May). The Social Psychology of Privacy. *American Journal of Sociology, 73,* 741–752.
16. Lombardi, D. B., & Ciceri, M. R. (2016). More Than Defense in Daily Experience of Privacy: The Functions of Privacy in Digital and Physical Environments. *Europe's Journal of Psychology, 12*(1), 115–136.
17. Oulasvirta, A., Pihlajamaa, A., Perkio, J., Ray, D., Vahakangas, T., Hasu, T., Vainio, N., & Myllymaki, T. (2012). Long-Term Effects of Ubiquitous

Surveillance in the Home. *Proceedings of the 2012 ACM Conference on Ubiquitous Computing*, 41–50. https://doi.org/10.1145/2370216.2370224.

18. Nissenbaum, H. (2010). *Privacy in Context: Technology, Policy, and the Integrity of Social Life*. Palo Alto, CA: Stanford University Press.

19. Jeff Smith, H., Dinev, T., & Xu, H. (2011, December). Information Privacy Research: An Interdisciplinary Review. *MIS Quarterly, 35*, 989–1015.

20. Solove, D. J. (2008). *Understanding Privacy*. Boston, MA: Harvard University Press.

21. Westin, A. F. (2003, July). Social and Political Dimensions of Privacy. *Journal of Social Issues, 59*, 431–453.

22. Dunbar, R. I. M. (2004). Gossip in Evolutionary Perspective. *Review of General Psychology, 8*(2), 100–110.

23. Dinev, T., McConnell, A. R., & Jeff Smith, H. (2015). Informing Privacy Research Through Information Systems, Psychology, and Behavioral Economics: Thinking Outside the 'APCO' Box. *Information Systems Research, 26*(4), 639–655.

24. Martin, K. D., & Murphy, P. E. (2017, March). The Role of Data Privacy in Marketing. *Journal of the Academy of Marketing Science, 45*, 135–155.

25. Milne, G. R., & Bahl, S. (2010). Are There Differences Between Consumers' and Marketers' Privacy Expectations? A Segment- and Technology-Level Analysis. *Journal of Public Policy & Marketing, 29*(Spring), 138–149.

26. Xu, H., Teo, H.-H., Tan, B. C., & Agarwal, R. (2012, December). Effects of Individual Self-Protection, Industry Self-Regulation, and Government Regulation on Privacy Concerns: A Study of Location-Based Services. *Information Systems Research, 23*, 1342–1363.

27. Bowie, N. E., & Jamal, K. (2006, July). Privacy Rights on the Internet: Self-Regulation or Government Regulation. *Business Ethics Quarterly, 16*, 323–342.

28. Martin, K. (2015). Privacy Notices as Tabula Rasa: An Empirical Investigation into How complying with a Privacy Notice Is Related to Meeting Privacy Expectations Online. *Journal of Public Policy & Marketing, 34*(Fall), 210–227.

29. Vail, M. W., Earp, J. B., & Antón, A. I. (2008, August). An Empirical Study of Consumer Perceptions and Comprehension of Web Site Privacy Policies. *IEEE Transactions on Engineering Management, 55*, 442–454.

30. Bansal, G., Zahedi, F. M., & Gefen, D. (2016). Do Context and Personality Matter? Trust and Privacy Concerns in Disclosing Private Information Online. *Information & Management, 53*, 1–21.

31. Winter, S. J., Stylainou, A. C., & Giacolone, R. A. (2004, October). Individual Differences in the Acceptability of Unethical Information Technology Practices: The Case of Machiavellianism and Ethical Ideology. *Journal of Business Ethics, 54*, 273–301.

32. Baruh, L., Secinti, E., & Cemalcilar, Z. (2017). Online Privacy Concerns and Privacy Management: A Meta-Analytical Review. *Journal of Communication, 67*, 26–53.

33. Goldfarb, A., & Tucker, C. (2012, May). Shifts in Privacy Concerns. *American Economic Review, 102*, 349–353.
34. Rainie, L., & Duggan, M. (2016). Privacy and Information Sharing. *Pew Research Center.* www.pewresearch.org.
35. Anderson, M., & Perrin, A. (2017). Tech Adoption Climbs Among Older Adults. *Pew Research Center.* www.pewinternet.org.
36. Phelps, J., Nowak, G., & Ferrell, E. (2000). Privacy Concerns and Consumer Willingness to Provide Personal Information. *Journal of Public Policy & Marketing, 19*(Spring), 27–41.
37. Tucker, C. (2014, October). Social Networks, Personalized Advertising and Privacy Controls. *Journal of Marketing Research, 51*, 546–562.
38. White, T. B., Zahay, D. L., Thorbjørnsen, H., & Shavitt, S. (2008). Getting Too Personal: Reactance to Highly Personalized Email Solicitations. *Marketing Letters, 19*, 39–50.
39. Tsai, J. Y., Egelman, S., Cranor, L., & Acquisti, A. (2011). The Effect of Online Privacy Information on Purchasing Behavior: An Experimental Study. *Information Systems Research, 22*(2), 254–268.
40. Bart, Y., Shankar, V., Sultan, F., & Urban, G. L. (2005, October). Are the Drivers and Role of Online Trust the Same for All Web Sites and Consumers? A Large-Scale Exploratory Empirical Study. *Journal of Marketing, 69*, 133–152.
41. Schlosser, A. E., White, T. B., & Lloyd, S. M. (2006, April). Converting Web Site Visitors into Buyers: How Web Site Investment Increases Consumer Trusting Beliefs and Online Purchase Intentions. *Journal of Marketing, 70*, 133–148.
42. Miyazaki, A. D. (2008). Online Privacy and the Disclosure of Cookie Use: Effects on Consumer Trust and Anticipated Patronage. *Journal of Public Policy & Marketing, 27*(Spring), 19–33.
43. Wirtz, J., & Lwin, M. O. (2009). Regulatory Focus Theory, Trust, and Privacy Concern. *Journal of Service Research, 12*(2), 190–207.
44. Aguirre, E., Mahr, D., Grewal, D., de Ruyter, K., & Wetzels, M. (2015). Unraveling the Personalization Paradox: The Effect of Information Collection and Trust-Building Strategies on Online Advertisement Effectiveness. *Journal of Retailing, 91*(1), 34–49.
45. Mikolajczak, M., Pinon, N., Lane, L., de Timary, P., & Luminet, O. (2010, September). Oxytocin Not Only Increases Trust When Money Is at Stake, but Also When Confidential Information Is in the Balance. *Biological Psychology, 85*, 182–184.
46. Martin, K. D., Borah, A., & Palmatier, R. W. (2017, January) Data Privacy: Effects on Customer and Firm Performance. *Journal of Marketing, 81*, 36–58.
47. Smith, N. C., & Cooper-Martin, E. (1997, July). Ethics and Target Marketing: The Role of Product Harm and Consumer Vulnerability. *Journal of Marketing, 61*, 1–20.
48. Vertesi, J. (2014, May 1). My Experiment Opting Out of Big Data Made Me Look Like a Criminal. *Time.* Available at http://time.com/83200/privacy-internet-big-data-opt-out/. Accessed March 1, 2018.

49. Solove, D. J. (2003). Identity Theft, Privacy, and the Architecture of Vulnerability. *Hastings Law Journal, 54*(1227), 1–47.

50. Identity Theft Resource Center. (2017). Retrieved from https://www.idtheft-center.org/Data-Breaches/data-breaches.

51. Martin, K. D., Borah, A., & Palmatier, R. W. (2018, February 15). A Strong Privacy Policy Can Save Your Company Millions. *Harvard Business Review.* hbr.org..

52. Feinberg, M., Willer, R., Stellar, J., & Keltner, D. (2012). The Virtues of Gossip: Reputational Information Sharing as Prosocial Behavior. *Journal of Personality and Social Psychology, 102*(5), 1015–1030.

53. Foster, E. K. (2004). Research on Gossip: Taxonomy, Methods, and Future Directions. *Review of General Psychology, 8*(2), 78–99.

54. Emler, N. (1994). Gossip, Reputation and Social Adaption. In R. F. Goodman & A. Ben-Ze'ev (Eds.), *Good Gossip* (pp. 119–140). Lawrence: University of Kansas Press.

55. Beersma, B., & Van Kleef, G. A. (2012). Why People Gossip: An Empirical Analysis of Social Motives, Antecedents, and Consequences. *Journal of Applied Social Psychology, 42*(11), 2640–2670.

56. Mills, C. (2010). Experiencing Gossip: The Foundations for a Theory of Embedded Organizational Gossip. *Group and Organization Management, 35*(4), 371–390.

57. Baumeister, R., Zhang, L., & Vohs, K. D. (2004). Gossip as Cultural Learning. *Review of General Psychology, 8*(2), 111–121.

58. Leary, M. R., & Leder, S. (2009). The Nature of Hurt Feelings: Emotional Experience and Cognitive Appraisals. In A. L. Vangelisti (Ed.), *Feeling Hurt in Close Relationships* (pp. 15–33). New York: Cambridge University Press.

59. Richman, L. S., & Leary, M. R. (2009). Reactions to Discrimination, Stigmatization, Ostracism, and Other Forms of Interpersonal Rejection: A Multimotive Model. *Psychological Review, 116*(2), 365–383.

60. Turner, M. M., Mazur, M. A., Wendel, N., & Winslow, R. (2003, June). Relational Ruin or Social Glue? The Joint Effect of Relationship Type and Gossip Valence on Liking, Trust, and Expertise. *Communication Monographs, 70*, 129–141.

61. Baker, S. M., Gentry, J. W., & Rittenburg, T. L. (2005, December). Building Understanding of the Domain of Consumer Vulnerability. *Journal of Macromarketing, 25*, 128–139.

62. Eder, D., & Enke, J. L. (1991, August). The Structure of Gossip: Opportunities and Constrains on Collective Expression Among Adolescents. *American Sociological Review, 56*, 494–508.

63. Cumbley, R., & Church, P. (2013, October). Is 'Big Data' Creepy? *Computer Law & Security Review, 29*, 601–609.

64. Williams, K. D. (2007). Ostracism. *Annual Review of Psychology, 58*, 425–452.

65. Kumar, V., Zhang, X., & Luo, A. (2014, August). Modeling Customer Opt-In and Opt-Out in a Permission-Based Marketing Context. *Journal of Marketing Research, 51*, 403–419.
66. Steel, E., & Fowler, G. A. (2010). Facebook in Privacy Breach. *Wall Street Journal, 256*(92), A1–A2.
67. McKinnon, J., & Hagey, K. (2018). As Mark Zuckerberg Prepares to Testify, Here's How Washington Could Regulate Silicon Valley. *Wall Street Journal.* Available at https://www.wsj.com/articles/as-zuckerberg-prepares-to-testify-questions-grow-over-how-to-protect-data-1523266201. Accessed May 22, 2018.

3

The Changing Privacy Legal Landscape

Introduction: The "Right" to Privacy

The term the "right to privacy" implies multiple freedoms, protected by laws, social conventions, and morals, including

- **Physical privacy**, or freedom from intrusion into one's physical space or solitude.
- **Associational privacy**, the freedom to form and maintain social and political groups.
- **Decisional privacy**, which refers to the freedom to stop others from meddling in intimate choices like birth control, abortion, marriage, medical care, and consensual adult sexual relationships.
- **Proprietary privacy**, which provides exclusive ownership and control over attributes of personal identity such as voice, name, and photographic likeness.
- **Intellectual privacy**, or the freedom to think about whatever you want.
- **Informational privacy**, which limits others' access to information that people consider sensitive or confidential.

This chapter centers on informational privacy and the rules that give people control over their personal data—defined as any information that can be used by itself or in combination with other information to identify someone.

Glory Francke, a US and EU data privacy regulatory lawyer, authored this chapter based on her experiences practicing privacy law in the European Union and United States.

The European Union (EU), Canada, and Japan each protect informational privacy using a single, comprehensive privacy law. But the United States takes a different approach and uses a hodgepodge of federal and state laws to protect informational privacy. Some regulate specific categories of information (financial, health care, medical), while others apply to usage activities (e.g., telephone marketing, text messaging, emailing). Hundreds of additional data protection laws exist at state levels, California alone has more than 25 unique privacy and data security laws on its books.

A review of privacy laws might seem to make for dull reading, but do not be fooled. Privacy law is emotional. It reflects excitement over technological discoveries, followed by indignation and outrage over invasions of privacy. The goal of this chapter thus is twofold. First, we seek to familiarize marketers with the legal landscape for informational privacy law. Second, we build on the psychology of privacy detailed in Chapter 2 to delineate the emotions that have led to the current sets of laws. That is, we recognize that legal compliance is mandatory, but we also acknowledge that understanding the emotional drivers of these legal requirements can help marketers protect their brands and create competitive advantages. A final note: The laws and regulations summarized in this chapter are complex and detailed. This information should not be taken as a guideline for action in specific situations. Rather, it provides an overview of oft-cited privacy law to heighten marketers' sensitivity to the legal requirements and sanctions that must be part of any successful marketer's consideration set.

The Patchwork of U.S. Privacy Protection

This generation is not the first to assert that technology is threatening privacy. That idea was originally promoted by Samuel Warren and Louis Brandeis, the founders of legal privacy scholarship, whose definition we provided in Chapter 1. Their 1890 article, "The Right to Privacy," is among the most cited *Harvard Law Review* publications ever. It flags the threat to privacy posed by social developments (tabloid newspapers) and new technology (portable cameras) [1]. Increased reporting on scandals and gossip drove newspaper circulation upward by about 1000% in the final decades of the 1800s [2]. At the same time, technological developments by the Eastman Kodak Company led to cameras that were portable enough to document public events [3]. Warren and Brandeis warned that this combination posed an imminent danger, such that "Instantaneous photographs and newspaper enterprises have invaded the sacred precincts of private and domestic life

(p. 195)" [4]. They called for the law to evolve and respond to these technological changes that threatened the right "to be let alone" [5]. Traditional prohibitions against trespass, assault, libel, and other invasive acts could not, in their view, protect individuals. A new legal regime was needed to enforce clear boundaries between public and private life and uphold the "right to one's personality" against threats from modern business practices and invasive inventions [6].

Wiretapping and (No) Constitutional Right to Privacy

Wiretapping to eavesdrop on private conversations developed in parallel with the adoption of the telephone. Even as Brandeis and Warren were warning against invasive inventions, the New York City Police Department began wiretapping initiatives [7]. State legislators reacted to such practices by passing a hodgepodge of legislation to prohibit wiretapping [8]. But it was not until shortly before Herbert Hoover became the first U.S. president to install a telephone in the Oval Office that the Supreme Court weighed in on whether wiretapping violated citizens' constitutional rights.

In *Olmstead v. United States*, Roy Olmstead, a Seattle bootlegger, argued that the method for gathering the evidence used to convict him—a wiretap in his home—violated his Fourth and Fifth Amendment rights [9]. The Fourth Amendment protects people's right to freedom from unreasonable intrusions ("searches and seizures") by the government. The Fifth Amendment right at issue was the right to protect oneself from self-incrimination. The Court disagreed with Olmstead's arguments. It held that wiretapping and listening to Olmstead's conversations did not constitute a "search," because the conversations were not material objects, like "papers" or "effects," so the officers committed no trespass on Olmstead's property when they tapped the public telephone wires [10]. Nor did the use of wiretapped conversations as incriminating evidence violate the Fifth Amendment protection against self-incrimination, because the conversations were conducted voluntarily.

Reasonable Expectation of Privacy

The Court's stingy interpretation of the Fourth Amendment in *Olmstead* led to widespread government eavesdropping and recording during the middle decades of the twentieth century. From 1941 to the mid-1960s, the FBI recorded almost a half million conversations in its attempt to monitor political groups, influence judicial appointments, threaten civil rights leaders, and intimidate or discredit members of Congress [11].

Then *Katz v. United States* came before the Supreme Court in 1967 [12]. It overturned its previous *Olmstead* ruling limiting Fourth Amendment protection to only certain areas or tangible objects. Rather, it found that government agents had conducted an illegal search when they placed a recording device outside a public telephone booth and recorded a telephone conversation [13]. The *Katz* case produced the two-party constitutional privacy test still in use today. According to the Katz test, a person has a "reasonable expectation of privacy" if the following two questions can be answered in the affirmative:

1. Did the person in question show an actual (subjective) expectation of privacy?
2. Does society recognize that expectation as "reasonable"?

Although the Katz test is "simple and concise on the page … its application is frequently puzzling, and its true nature remains something of a mystery" [14]. Although *Katz* establishes that citizens have a reasonable expectation of privacy against searches and seizures by police, it is unclear if, when, and how that right can be reliably invoked. In short, Constitutional law is not a clear or reliable source of individual privacy protections. People looking for a legal basis for a right to privacy instead must turn to federal and state laws and policies, founded in large part on Fair Information Practices (FIPS).

Federal Privacy Laws

Fair Information Practices

The development of the computer in 1946 revolutionized the way records and data were collected, disseminated, and used. It also raised considerable public concerns about privacy and catalyzed the need for specific data privacy protection laws. With digitization, vast amounts of personal data moved from ink and paper to a series of ones and zeroes, making it possible to replicate, compress, disseminate, analyze, or organize personal data on a massive, computerized scale. These factors also increased identity theft and fraud, because the personal data needed to commit various types of identity theft moved back and forth across computers regularly.

Digitization also led government agencies in the United States, Canada, and Europe to study so-called information practices more closely, to determine how entities should be allowed to collect and use personal information. The result was a series of reports, guidelines, and model codes that now represent widely accepted principles called the Fair Information Practices (FIPs). The five core FIPs are (1) notice/awareness, (2) choice/consent, (3) access/participation, (4) integrity/security, and (5) enforcement/redress.

US privacy law generally relies on notice and choice aspects, with the theory that when a website visitor arrives at a new homepage, the first thing she or he does is read the privacy policy (notice) and decide whether it adequately protects personal data (choice).

In particular, U.S. privacy law generally relies on notice and choice aspects, with the theory that when a website visitor arrives at a new homepage, the first thing she or he does is read the privacy policy (notice) and decide whether it adequately protects personal data (choice). There are obvious problems with this theory. In reality, who reads privacy policies? Is it even possible to read them all? One study calculated that the average person would need to spend 40 minutes a day reading privacy policies if they were to read a privacy policy each time they visited a new site [15]. And other studies show that even if everyone read all the policies, consensus on their content is unlikely, because they often are so ambiguously worded that they cause confusion among laypersons and experts alike [16]. Yet privacy policies remain a key factor in establishing the lawfulness of digital marketing efforts. Running afoul of U.S. data privacy law often results from firms' violations of their own privacy policies and the data privacy and security promises they contain.

The Fair Credit Reporting Act

The first federal data privacy law was the Fair Credit Reporting Act of 1970 (FCRA) [17]. It was triggered by outrage over how consumer reporting agencies (CRAs) gathered and sold information about people's creditworthiness [18]. "Investigative" reports gathered information about people's sexual orientation and activity, alcohol consumption, and gossip from neighbors and friends, such as rumors about police involvement. Companies then used this information to determine whether to offer services to the person.

A representative from the automobile insurance industry argued that CRAs helped weed out potential consumers with "deviant behavior characteristics," such that they "wear pink shorts, or have long hair and a mustache,... read Karl Marx"—behavior that allegedly would make them difficult to defend before a jury [19].

The complex FCRA has undergone multiple amendments. It regulates credit reports, consumer investigatory reports, and employment background checks; it gives people the right to access and correct their personal data. It also imposes stringent requirements related to data usage, destruction, notice, consent, and accountability. All CRAs also must follow reasonable procedures to protect the confidentiality, accuracy, and relevance of credit information [17].

A company constitutes a CRA under the law if it assembles and evaluates consumer report information—on a person's character, reputation, or personal characteristics—for the use of third parties. Such reports tend to be prominent in employment, housing, or credit markets, and their use triggers a company's FCRA obligations. Failures to implement any of the accuracy, dispute, or other safeguards required by the law could lead to harm to people's reputations, employment prospects, and ability to access credit.

And a company cannot avoid FCRA obligations by issuing disclaimers that its reports are not FCRA-complaint or should not be considered screening products. The Federal Trade Commission (FTC) discounts such disclaimers and considers instead whether the business should reasonably believe that its information is being utilized for employment or other FCRA purposes. In one FTC case, a company called Filiquarian relied on a FCRA disclaimer but also specifically advertised that its reports could be used to inform hiring decisions, promising prospective clients: "Are you hiring somebody and wanting to quickly find out if they have a record? Then Texas Criminal Record Search is the perfect application for you" [20]. According to the FTC, Filiquarian's reports clearly qualified as consumer reports under the FCRA. Filiquarian also violated the FCRA by failing to maintain reasonable procedures to verify users' identities or guarantee that the information would be used for a permissible purpose; lacking procedures to ensure that the information it provided in consumer reports was accurate; and failing to notify users of the information included in the consumer reports. The takeaway for marketers? Companies offering background screening products for employment or other FCRA purposes, as well as the businesses that rely on them, absolutely must stay on the right side of the law by paying close attention to what the FTC says, including in its warnings to other companies.

Gramm-Leach-Bliley Act (GLBA)

Congress enacted the Gramm-Leach-Bliley Act (GLBA) in 1999 to modernize the financial service industry and allow banks, stock brokerage houses, and insurance companies to merge. Such mergers would give the newly created entities access to large swaths of personal data. Some members of Congress sought to include privacy protection in the GLBA by introducing an amendment (the "Markey Amendment"). But the banking industry lobbied hard against this Amendment, especially provisions that would have given consumers notice and choice about how their information would be shared.

But critical support for the Markey Amendment came from an unexpected source on a personal mission. Texas-based Republican Representative Joe Barton maintained a Washington, D.C. residence, where he stayed while Congress was in session. Victoria's Secret catalogs addressed to Rep. Barton began arriving at his Washington address. The representative was horrified. How did Victoria's Secret get his address? Neither he nor his wife ever shopped there. And worse, what would his wife think? Suspecting that his Washington credit union sold his address to Victoria's Secret, Barton supported the Markey Amendment. And with this bipartisan support, the Markey Amendment became part of the GLBA, ensuring that individual consumers have the right to direct financial institutions not to sell their personal information to third parties [21]. A financial institution is any company that offers financial products or services, like loans, financial or investment advice, or insurance, including banks, non-bank mortgage lenders, real estate appraisers, loan brokers, financial or investment advisers, debt collectors, tax return preparers, and real estate settlement service providers. Each of these financial institutions must provide its customers the right to opt out before it may share any of their protected customer information with a third party.

Medical Privacy Law for Marketers

Personal health information is protected by the Health Insurance Portability and Accountability Act (HIPAA), which encompasses the HIPAA Privacy Rule [22] and the HIPAA Security Rule [23]. This Act regulates "covered entities," including healthcare providers, plans, and clearinghouses, as well as "business associates," which are entities that contract with covered entities and receive, use, and process information from them. The Privacy Rule governs protected health information (PHI), defined broadly as any "individually identifiable health information," which includes paper records. The Security Rule is narrower and regulates only electronic PHI (ePHI). Furthermore, HIPAA imposes important workforce training requirements. In any covered entity, HIPAA compliance is everyone's responsibility, and HIPAA violations can come from any department, not just direct care providers. Exhibit 3.1 provides an example of how managers can get in trouble with HIPAA as well as some key lessons learned.

Exhibit 3.1: HIPAA: Lessons Learned [24]

A not-for-profit healthcare system recently agreed to a $2.4 million settlement over the alleged disclosure of one patient's identity without her consent. The patient in question checked in for a follow-up visit with her gynecologist. During the check-in process, a clinic staff member thought the patient's driver's license looked suspicious. Consequently, the office called the licensing bureau of the Texas Department of Public Safety (DPS), which then instructed the office to contact local law enforcement. After confirming the false license number, local law enforcement made a decision to arrest the patient.

The clinic had complied with HIPAA up to this point: The Privacy Rule allows providers to report PHI (including driver's license information) if it relates to evidence of a crime that has occurred on the entity's premises. But the arrest also sparked protests. The patient was an undocumented immigrant who had health insurance under her husband's private plan. Her crying, 8-year-old, U.S.-born daughter witnessed the arrest. Immigrant advocates questioned whether the arrest would have a chilling effect on other undocumented immigrants seeking medical care. The healthcare system responded with a press release, calling the incident "unfortunate" and stating "quality and safety reasons" for the process that led to the call to DPS. But the press release also named the patient.

About two months later, the U.S. Department of Health and Human Services initiated a compliance review of the healthcare system, responding to multiple reports alleging the disclosure of the patient's PHI to the media and various public officials, without the patient's authorization. Some key lessons for marketers emerged from this incident:

- *If you think it might be PHI, it probably is.* Regulators interpret PHI broadly to include any information that identifies someone as a patient. When in doubt, leave information out.
- *Public knowledge is no excuse.* Even if someone (e.g., the media) knows an individual was a patient, the provider cannot release additional PHI or even confirm that the individual was a patient without a valid basis under HIPAA.
- *HIPAA protects everyone.* Every patient's PHI is protected, regardless of immigration status or criminal acts, even if the act was committed on the covered entity's premises.
- *HIPAA compliance is everyone's job.* A HIPAA violation can result from statements by public affairs or government relations departments, as well as from leaders or care providers. Again, when in doubt, leave information out.

Commercial Communications and Privacy

Some privacy laws regulate the collection and use of personal data. Privacy laws that regulate commercial communications differ though. These "anti-marketing" laws primarily regulate how businesses may contact a person. They impose specific requirements, liability, and even criminal sanctions for some violations [25]. The details vary, depending on the communication medium. Exhibit 3.2 provides a summary of the most critical US Federal Privacy Laws.

Email. Congress tried to control the problem of unwanted, unsolicited emails by adopting a federal law whose name was inspired by a Monty Python skit: Controlling the Assault of Non-Solicited Pornography and Marketing Act (CAN-SPAM Act, 2003). In the famous comedy skit, a restaurant customer tries to order a SPAM-free entrée, only to find that SPAM is in literally everything on the menu [26]. His protests, "But I don't like SPAM," are drowned out by the other restaurant goers (a group of Vikings, of course), who break into song praising the meat-like substance every time they hear the word. Early chat room users coopted the term "spam" to refer to undesirable emails [27]. And just as SPAM singing drowned out conversation about the undesirability of SPAM itself, even though it popped up all over the menu, so too were unsolicited emails unwanted, impossible to avoid, and difficult to discuss reasonably above the noise of pornographic content or get-rich-quick schemes.

The CAN-SPAM Act gives commercial email recipients the right to block future emails (opt-out) and creates penalties for violations. It applies to "commercial electronic mail messages," or emails whose primary purpose is a commercial advertisement or promotion of a commercial product or service [28]. It does not matter if the emails are sent in bulk or to just one person, so long as they meet that definition. Each email violating the CAN-SPAM Act is subject to penalties of up to $40,654, hence non-compliance can be costly. Following this law is not complicated: Don't send an email without checking to ensure it complies with CAN-SPAM.

Telephone Communications. Imagine a beloved senior member of your family was interrupted one Sunday morning by a phone call from a stranger, wishing to sell them burial arrangements. Most people would consider this call an indefensible intrusion, the kind of thing that should be illegal. Spurred by such real-life scenarios, Congress enacted the Telephone Consumer Protection Act (TCPA) in 1991 and authorized the Federal Communications Commission (FCC) to implement the related rules and

Exhibit 3.2 Selected U.S. Federal Privacy Laws

Selected Federal Privacy Laws			
Law	What it protects	What it regulates	Regulator
Federal Trade Commission Act (15 U.S.C. §§41–58) (FTC Act)	Consumers' personal data	Unfair or deceptive trade practices., such as failing to comply with posted privacy policies or unauthorized disclosures of personal data	FTC
Children's Online Privacy Protection Act (COPPA) (15 U.S.C. §§6501–6506)	Children's personal data	Online collection of information from children	FTC
Financial Services Modernization Act (Gramm-Leach-Bliley Act [GLBA]) (15 U.S.C. §§6801–6827)	Nonpublic personal information, which is any personally identifiable financial information that is not publicly available and is collected for the purpose of providing a financial product or service	The collection, use, and disclosure of financial information	FTC
Health Insurance Portability and Accountability Act (HIPAA) (42 U.S.C. §1301 et seq.)	Protected health information (PHI)	The collection and use of PHI by healthcare providers, data processors, pharmacies, and other entities that come in contact with medical information	HHS's Office of Civil Rights
Fair Credit Reporting Act (FCRA) (15 U.S.C. §1681) (also; Fair and Accurate Credit Transactions Act, Pub. L. No. 108–159, which amended the FCRA)	Consumer reports, or any communication issued by a consumer reporting agency that relates to a consumer's creditworthiness, credit history, credit capacity, character, and general reputation, used to evaluate a consumer's eligibility for credit or insurance	Consumer reporting agencies, those that use consumer reports (e.g., lenders), and those that provide consumer reporting information (e.g., credit card company)	FTC

(continued)

Exhibit 3.2   (continued)

Selected Federal Privacy Laws

Law	What it protects	What it regulates	Regulator
Telephone Consumer Protection Act (TCPA) (47 U.S.C. §227 et seq.)	Individual telephone numbers	The collection and use of telephone numbers, to prevent for example unsolicited text message marketing	FCC and FTC
Controlling the Assault of Non-Solicited Pornography and Marketing Act (CAN-SPAM Act) (15 U.S.C. §§7701–7713 and 18 U.S.C. §1037)	Individual email addresses	The collection and use of e-mail addresses	FTC
Electronic Communications Privacy Act (18 U.S.C. §2510) and Computer Fraud and Abuse Act (18 U.S.C. §1030)	Individual electronic communications (including email)	The interception of electronic communications and computer tampering, respectively	FTC

Notes *FTC* = Federal Trade Commission, *HHS* = Department of Health and Human Services, *FCC* = Federal Communications Commission regulations [29]. The TCPA restricts how businesses may contact consumers' telephones, and it allows consumers to opt out of receiving such calls [30]. The provisions that trigger most of the litigation or enforcement actions are prohibitions against:

- Calling cell phone numbers using automatic telephone dialing systems (autodialers or ATDSs) or an artificial or prerecorded voice without consumer consent.
- Calling residential telephones using an artificial or prerecorded voice to deliver a message without:

  - prior express consent (if for a commercial purpose); or
  - appropriate disclosure language.

- Sending unsolicited fax advertisements without appropriate consent or opt-out disclosures.
- Making marketing calls to residential consumers listed on the National Do-Not-Call Registry.

The TCPA also gives individual citizens the right to bring a lawsuit against abuses (i.e., private right of action). The costs for breaking the law can add up quickly; courts have awarded damages of as much as $500–$1500 for each text, per plaintiff. Unwanted calls make up the largest category of complaints the FCC receives, including more than 215,000 complaints in just one recent year [31]. Here the takeaway for marketers again is clear: Make telephone-based marketing campaigns comply with TCPA, to avoid costly actions that annoy potential consumers.

Direct Mail. Direct mail communications, sent by the postal service, are highly regulated, though the rules apply uniquely to different categories of postal communications. Marketers can consult the Domestic Mail Manual for the Mailing Standards of the U.S. Postal Service, which clearly establishes their responsibilities, before sending any marketing communications by direct mail [32].

Children's Personal Data and Marketing

As the Internet grew during the 1990s, marketers began to collect personal information from children who registered with chat rooms and discussion boards, tracking their web surfing behavior and promising rewards and gifts in exchange for their personal information. The compiled information was sold to third parties, but many advertisers used disingenuous methods to obtain this valuable information [33]. For example, a D.C. Comics website asked children to provide personal data with the exhortation, "Good citizens of the Web, help commissioner Gordon with the Gotham Census (p. 195)" [34]. News also began circulating about child predators using the Internet to find children [35]. A tipping point came when a CNN reporter used the name of a notorious, real-life child kidnapper and murderer to purchase personal information, including names and addresses, of 5500 children from a data broker [36].

Congress responded to the public outcry by adopting the Children's Online Privacy Protection Act of 1998 (COPPA). By giving parents control over the information gathered about their children online, COPPA's primary goal is to put parents in a position of responsibility. This law applies to operators of commercial websites and online services (including mobile apps) that target children under the age of 13 years and that collect, use, or disclose their personal information, as well as operators of general audience websites that know they are collecting, using, or disclosing personal information from children under 13 years of age [37].

Litigation, Regulatory, and Financial Risks of Data Breaches

As the seemingly constant reports of data breaches show, no industry is untouched, and no organization is entirely safe from the loss, theft, or compromise of consumers' personal data. Companies are expected to exercise reasonable care to protect the personal information of their customers and employees, so sophisticated companies hope for the best while preparing for the worst—whether litigation, regulatory fines, or high out-of-pocket response costs, along with the unquantifiable damage to their reputation.

State-by-State Legal Variations

As may come as no surprise, the U.S. law features 50 different state-level data breach notification laws [38]. Their contents vary, but in general, notification obligations are triggered when unauthorized access (breach of security) includes personal information, as defined in each statute, which tends to include the person's first name or initial and last name, in combination with some other data element (e.g., Social Security number, driver's license number, financial account information with access code). Some state definitions of personal information are broader and include data elements such as biometric or health information, usernames, or email addresses and passwords. In states such as Illinois, Washington, and Texas, biometric privacy laws already are on the books, and several other states have proposed similar regulations. Other common state-level data breach requirements include [39]:

- Notification of affected state residents without unreasonable delay.
- Notification of certain agencies, including states' attorneys general or consumer agencies, under certain circumstances.
- Notification exceptions for good-faith access by an employee, encryption of the data, and determinations of low risk of harm.
- Specific requirements for the content of the notification.
- Civil penalties enforced by the state's attorney general.

In other areas, the laws diverge more significantly. For example, in some states, notification to state agencies is required only when a certain number of residents have been affected, whereas other states demand notification to state agencies regardless of how many were affected. All states require notification "without unreasonable delay," but only some of them state a specific duration (e.g., within 30, 45, or 60 days). Some states allow only the attorney general to bring a lawsuit for violations, but others support a private right of action. Therefore, multistate marketers must be familiar with each variation (see Exhibit 3.3 for summary of state data breach notification laws).

Regulatory Fines and Financial Risk

State data breach laws impose notification and follow-up costs, associated with creating contact databases, determining all regulatory requirements, engaging outside experts, postal expenditures, email bounce-backs, and inbound communications. Response costs also include help desk activities, inbound communications, special investigative activities, remediation, legal expenditures, product discounts, identity protection services, and regulatory interventions. Each year, IBM Security and the Ponemon Institute release a widely read report detailing the average costs of a data breach for a company [40]. According to the 2017 report, data breaches cost U.S. organizations an average of $225 per record, though strictly regulated industries suffer higher data breach costs; healthcare foots the largest average bill, at $380 per record [41].

Exhibit 3.3   State Data Breach Notification Laws

Issue	States
Notification content requirements	Alabama, California, Florida, Hawaii, Illinois, Iowa, Maryland, Massachusetts, Michigan, Missouri, Montana, New Hampshire, New Mexico, New York, North Carolina, Oregon, Rhode Island, South Carolina, Vermont, Virginia, Washington, West Virginia, Wyoming, and Puerto Rico
Broad definition of "personal information"	Alabama, Alaska, California, Connecticut, Delaware, Florida, Georgia, Illinois, Iowa, Kansas, Maine, Maryland, Massachusetts, Missouri, Montana, Nebraska, Nevada, New Hampshire, New Jersey, New Mexico, New York, North Carolina, North Dakota, Ohio, Oregon, Rhode Island, South Carolina, South Dakota, Texas, Vermont, Virginia, Wisconsin, Wyoming, District of Columbia, and Puerto Rico
Notification to state agency required	Alabama, Alaska, California, Connecticut, Delaware, Florida, Hawaii, Idaho, Illinois, Indiana, Iowa, Louisiana, Maine, Maryland, Massachusetts, Missouri, Montana, Nebraska, New Hampshire, New Jersey, New Mexico, New York, North Carolina, North Dakota, Ohio, Oregon, Rhode Island, South Carolina, South Dakota, Texas, Vermont, Virginia, Washington, Wisconsin, and Puerto Rico
Credit monitoring required	California, Connecticut, and Delaware
Risk of harm notification waiver	Alabama, Alaska, Arizona, Arkansas, Colorado, Connecticut, Delaware, Florida, Hawaii, Idaho, Indiana, Iowa, Kansas, Kentucky, Louisiana, Maine, Maryland, Massachusetts, Michigan, Mississippi, Missouri, Montana, Nebraska, Nevada, New Hampshire, New Jersey, New Mexico, New York, North Carolina, Ohio, Oklahoma, Oregon, Pennsylvania, South Carolina, South Dakota, Tennessee, Utah, Vermont, Virginia, Washington, West Virginia, Wisconsin, and Wyoming

Litigation

Plaintiffs have attempted to recover money for medical bills, pain and suffering, and diminished quality of life ("damages") by suing companies responsible for data breaches that are alleged to have caused physical, mental, or emotional injury ("harm"). Such a lawsuit must give a specific legal reason the court should award damages, whether because the company broke a specific provision of a law, violated a contractual promise, or exhibited negligence (collectively, "legal theories"). Most data breach lawsuits rely on the legal theory of negligence; most of them also fail. It is difficult to establish sufficient consumer harm, shame, humiliation, or financial loss stemming from an isolated invasion of privacy [42]. Individual consumers have argued unsuccessfully that a violation of a privacy policy is akin to breach of a contract [43].

Considering the difficulty that plaintiffs face in proving that they have been injured by a data breach, which would give them standing to bring a lawsuit, the risk of litigation for firms has remained relatively low. When it comes to informational privacy, consumers instead rely on the FTC to enforce the consumer protection provisions of the FTC Act and federal privacy protection laws.

The Privacy Regulator: Federal Trade Commission

The scattershot U.S. approach to data protection leaves several areas seemingly unregulated. There are no federal laws governing how companies like Facebook or Google may collect and use personal data. E-commerce companies like Amazon collect and use personal data without any directly applicable federal rules too. But despite the absence of a comprehensive data protection law, consumer privacy (at least theoretically) is protected by the Federal Trade Commission's power to prevent "unfair or deceptive acts or practices in commerce" under Section 5 of the Federal Trade Commission Act (FTC Act).

Initially created to "bust the trusts," Congress expanded the FTC's jurisdiction to consumer protection. Section 5 gives the FTC two distinct sources of authority to bring an enforcement action against a company: (1) deception and (2) unfairness. Section 5 also establishes the FTC's informal role as an exhorter of best practices for digital ecosystems. Many FTC enforcement actions feature a legal double-dip, such that they cite violations of Section 5 together with violations of another privacy statute over which the FTC has authority, such as FCRA or COPPA.

> There are no federal laws governing how companies like Facebook or Google may collect and use personal data. Ecommerce companies like Amazon collect and use personal data without any directly applicable federal rules too.

Mechanics of an FTC Enforcement Action

Investigation. The FTC Act gives the FTC broad power to gather and compile information about most business practices that affects commerce [44]. Some investigations are informal. For example, the FTC hosts industry workshops on new technologies and sends unenforceable requests for information through so-called access letters. But if a company declines to respond, the FTC can move to a formal process and issue a legally enforceable demand for documents and written answers to questions ("Civil Investigative Demand," or CID). The FTC uses CIDs to investigate suspected violations. A federal court can compel a company to respond with written reports or answers to questions.

Proposed Complaint, Settlement, and Consent Orders. Staff attorneys analyze CID responses and, if merited, issue a complaint describing the alleged act(s) and potential remedy [45]. The business under investigation ("Respondent") has two options: (1) dispute the charges by going to court (i.e., litigating) or (2) enter into a settlement agreement with the FTC [46]. In a settlement agreement, the company does not concede guilt but agrees to waive all rights to judicial review and to accept an FTC consent order that details how it must remediate the matter. Consent orders commonly include financial penalties, prohibitions of specific activities, and corrective action and oversight requirements that can last for up to 20 years. For example, they might require the implementation of comprehensive privacy and security programs, biennial assessments by independent experts, monetary redress for consumers, the return of ill-gotten gains, deletion of illegally obtained consumer information, or robust transparency and choice mechanisms for consumers.

Despite the stringency of their commitments under the settlement agreements, almost all companies choose to settle: to avoid conceding any wrongdoing, to limit monetary penalties (because the courts can impose steeper fines than the FTC), to avoid the expense and uncertainty of litigation, and to mitigate concerns about judicial deference to the FTC [47]. Courtroom challenges are so infrequent that there are only a handful of judicial opinions available [48].

What Marketers Can Learn from FTC Consent Orders

Although FTC consent orders lack any binding precedential value—the FTC is not required to handle similar matters in the same fashion—it tends to be consistent. Therefore, consent orders grant marketers an excellent source of practical guidance regarding lawful digital marketing. Privacy lawyers parse every word to glean practical insights [49]. The consent orders and corresponding complaints are posted on the FTC's website and help define the legal boundaries for permitted and impermissible uses of personal data, as listed in the following subsections.

Don't Be Deceptive. The FTC historically has brought privacy cases on the basis of the deception prong of the FTC Act's Section 5. A trade practice is deceptive if it [50]:

1. Is an act (representation, omission, or practice),
2. Is likely to deceive a reasonable consumer, and
3. Results in a material loss to the consumer.

Marketers also can learn several additional lessons from FTC enforcement actions based on the deception prong of Section 5.

Keep (Explicit and Implied) Promises. The FTC initially protected personal data by bringing enforcement actions against companies that violated their own published privacy policies. In particular, the FTC has entered into consent agreements with companies for breaking promises, including promises to maintain confidentiality, refrain from disclosing information to third parties [50], clarify which data they collect [51], provide adequate security for personal data [52], keep data anonymous [53], or related to selling data [54]. Over time, the FTC has moved from enforcing such explicit company promises set out in policies to enforcing more implicit promises, including those that appeared only fleetingly on a website. Exhibit 3.4 provides a case example of Uber's broken data privacy promises and the resultant FTC complaint.

Exhibit 3.4: Uber's Broken Promises Case

The FTC's 2017 settlement with Uber demonstrates how a company might fail to meet promises made in statements that periodically appeared on its own website [55]. Uber's mobile apps collect sensitive personal information about both drivers (name, email address, phone number, postal address, profile picture, Social Security number, driver's license information, bank routing and account information, vehicle registration and insurance information) and passengers (names, email addresses, postal addresses, profile pictures, and detailed trip records including precise geolocation information). In late 2014, popular press articles began reporting on Uber's tendency to host lavish parties and entertain partygoers with its "God's View" tool, which showed silhouettes of users waiting for their rides and their locations; it also shared the "Creepy Stalker View" that tracked users' whereabouts in real time [56].

The public was not amused. Posts with titles like "Can We Trust Uber?" went viral [57]. Uber responded with a statement on its website, claiming to have implemented "a strict policy prohibiting all employees at every level from accessing a rider or driver's data," along with assurances that it was "closely" monitoring and auditing this policy [55]. But according to the FTC's complaint, Uber failed to meet these promises. It did not develop a system for ongoing monitoring of employee access to personal data. It systematically ignored automated alerts of potential misuses of personal information. Furthermore, the only accounts granted extra monitoring protections were those of internal, high-profile accounts, such as Uber executives.

The FTC's complaint also alleged that Uber made and broke data security promises. Similar FTC actions against unfair business practices related to data privacy and security have slowly increased. In this case, Uber issued a privacy policy that expressly applied to its websites and apps and contained specific security promises. According to the FTC, Uber broke these security promises in myriad ways; together, they represent a sort of hall of shame of what *not* to do when it comes to data security. In particular, Uber

- Permitted programs and engineers to use a single access key, granting them full administrative privileges over all data available in Uber's servers.
- Failed to restrict access to systems according to employees' job functions.
- Did not require multifactor authentication for access to its servers.
- Stored sensitive personal information on its cloud server in clear, readable text rather than encrypting the information—akin to leaving the vault and bank door wide open and unguarded [55].

In summary, Uber failed to implement even the most basic of security configurations, including on third-party servers storing its consumers' and drivers' personal information.

Be Honest. Not all deceptive acts involve a breach of a privacy promise. Deceptive actions also might trick consumers into disclosing their personal information. For example, in *FTC v. Craig Brittain*, the FTC alleged that Brittain obtained photographs by posing as a woman on Craigslist and, after sending nude photographs purportedly of himself (as a woman), soliciting nude photographs from other women in return [58].

Be Transparent/Give Notice. Many FTC complaints fault companies for failing to give customers sufficient notice about the collection and use of their personal data, or their choices for these uses [59]. In *FTC v. Lenovo, Inc.*, it charged the laptop manufacturer with a deceptive failure to disclose, noting that Lenovo had sold laptops with preinstalled "man-in-the-middle" software [60]. When a laptop user browsed the Internet, the software sat proverbially in the middle, routing and observing the traffic flowing between the laptop and the user's activity. This software could monitor sensitive data and then serve up ads tailored to the user's browsing history. As we emphasize throughout this book, when it comes to consumers' personal information, transparency and control represent gold standards. Lenovo was punished because it failed to tell consumers about the very existence of this software, nor did it solicit or receive consumer consent to allow the preinstalled software to intercept their Internet communications and browsing, then transmit those data to third parties.

Be Fair. The FTC started exercising its authority to go after unfair trade practices in 2003, and it has increasingly relied on that standard since [61]. The FTC's understanding of the unfairness doctrine reflects an evolutionary process, by which the standard has been refined over time to reflect decisions in cases, rules, and FTC statements. Thus, the FTC currently identifies unfair trade practices with a three-part test: A trade practice is unfair if it (1) causes substantial injury, (2) is not reasonably avoidable, and (3) is not balanced with countervailing benefits [62]. Although the FTC has yet to summarize its approach in a unified way, its actions reveal clear takeaways for marketers.

Don't Make Retroactive Changes to Privacy Policies. Retroactive changes to a privacy policy are unfair. For example, if a user submits data to a website whose privacy policy states, "We will not sell your data to third parties," that company cannot later remove that promise from its privacy policy and begin to sell the user's data. This scenario was the cause of one of the FTC's first unfair practice matter which involved Facebook [63]. Facebook changed its privacy policy—ostensibly to provide more privacy— but, as the FTC noted, in reality, it had reclassified certain user information as publicly available, such that users could no longer restrict access to their Friend Lists, Profile Picture, or Pages. Thus the FTC took action against the social media site, because the unfair changes exposed users' profiles to the public.

Don't Be Sneaky. Secretly collecting and selling consumer data is a nonviable business practice. Just ask Vizio, the smart television manufacturer that agreed to pay $2.2 million to settle charges that it preinstalled software on its products to collect viewing data from 11 million consumers' televisions, without their knowledge or consent [64]. According to the agency's complaint, Vizio's smart televisions could capture second-by-second information about the video displayed on the smart device, then integrate specific demographic information with these viewing data, such as the viewers' gender, age, income, and marital status. Allegedly Vizio then unfairly sold this valuable information to third parties that used it for various purposes, including targeting advertising to consumers through the devices.

European Data Protection Law

For many marketers, lawful digital marketing depends on compliance with global privacy laws. The most important of these is probably the European Union's General Data Protection Regulation (GDPR), due to its bar-raising rigor and extraterritorial reach. It took effect in May 2018 and applies to companies established in any of the European member states but also any company that processes the personal data of any EU resident, whether to offer goods or services that will be consumed in Europe or to monitor that resident's behavior in Europe.

The GDPR aims to regulate not just data collection but also data *correlation*. To that end, it defines personal data broadly as any piece of information, no matter how small, that can be combined with other pieces to create a picture that identifies an individual. Mobile device identifiers, IP addresses, and geolocation data are all personal data. Some information, such as insights that reveal political opinions, trade union membership, sexual orientation, genetics, or biometric data (among other things), are further classified as sensitive data, which require an even more elevated level of protection.

Legal Basis for Processing Personal Data

Under the GDPR's golden rule of data protection, a company cannot process a European resident's personal information without a legal basis. To be compliant, a privacy statement must identify a legal basis for each reason a company claims for processing personal data, whether (1) consent, (2) legitimate interest, (3) necessity for the performance of a contract, (4) legal compliance (with another law), (5) protecting vital interests, or (6) the public interest. Most marketers turn to consent or legitimate interest bases, though each poses a challenge. Consent requires a direct relationship with the data subject while a company's legitimate interest is only valid if it outweighs the data subject's right to privacy.

Because the GDPR took effect in May 2018, regulatory enforcement action and guidance is sparse thus far. But its predecessor, the Data Protection Directive (DPD), was based on the same principles, so actions taken against Facebook by European Data Protection Authorities (DPAs) just weeks before the GDPR came into effect likely foreshadow the GDPR's application in the future. Exhibit 3.5 provides an example of how Facebook got into regulatory troubles in Europe after its acquisition of WhatsApp when they changed their privacy policies regarding data sharing between the two entities.

Exhibit 3.5: WhatsApp: A Warm Welcome and a Cold (Regulatory) Shoulder [65]

In 2014 Facebook offered $19 billion in stock and cash to purchase the mobile messaging application WhatsApp. That price tag—one of the highest ever paid in the tech industry—suggests that Facebook welcomed WhatsApp to its family with open arms. European regulators did not share this enthusiasm. The acquisition and integration of WhatsApp into the Facebook universe already has prompted more than €100 million in regulatory fines, along with enforcement actions by several member state DPAs: the French Commission Nationale de l'Informatique et des Libertés, Spain's Agencia Española de Protección de Datos, and the United Kingdom's Information Commissioner's Office (ICO).

The WhatsApp founders Jan Koum and Brian Acton had long managed WhatsApp in line with their central mantra: "No ads! No games! No gimmicks!" After the Facebook acquisition, Koum published a blog post to calm WhatsApp users' privacy concerns, explaining: "Here's what will change for you, our users: nothing" [66]. Under its terms of service and privacy policy in force at that time, WhatsApp did not share user data with Facebook. When Facebook notified the European Commission of its acquisition, it also indicated it could not access those data operationally. According to statements from the European Commission, Facebook claimed it was "unable to establish reliable automated matching between Facebook and WhatsApp user accounts."

But regulatory trouble started when WhatsApp announced updates to its privacy policy and terms of service on August 25, 2016. The new policy gave users a choice: Stop using WhatsApp or accept that their personal information (but not message content) would be shared with the "Facebook family of companies" to support targeted advertising, among other things. The update was not well received. Within 24 hours, the ICO announced it would investigate the update. Other DPAs followed suit, and a coalition of Europe's national DPAs (called the Article 29 Working Group) notified WhatsApp of its serious concerns about the validity (or lack thereof) of its attempt to gain users' consent for the proposed data sharing.

According to ICO's summary, Facebook Ireland responded by halting data sharing for targeted advertising purposes by the end of October 2016 (while data sharing for security and safety purposes continued). Meanwhile, the scandal prompted questions among the European Commission about Facebook's statement that it could not reliably match Facebook and WhatsApp user accounts. In a press release issued in May 2017, it issued Facebook a fine of €110 million, for providing misleading information about the acquisition.

By March 2018, Facebook entered into an undertaking agreement with ICO in which it voluntarily committed to refrain from sharing any data between Facebook and WhatsApp until the GDPR came into force. After the GDPR came into effect, the two services may share data only for safety and security purposes, in full compliance with the GDPR. For advertising purposes, they may share data only in collaboration with and subject to the approval of the national DPA s.

Lessons to Learn for the GDPR

When regulators assessed a proposed data sharing agreement between WhatsApp and Facebook, the national laws stemming from the DPD identified it as illegal. Even as the GDPR has replaced the DPD, these regulatory analyses remain highly relevant. In a sense, the GDPR is the DPD on steroids; its requirements are more muscular, and compliance is more oner-ous, but the underlying data protection principles are the same. Companies thus can learn key lessons about GDPR compliance from the rejected Facebook–WhatsApp data sharing alliance.

Know the Legal Basis for Processing. According to the U.K. Information Commissioner's Office (ICO), the proposed processing agreement would be illegal under the DPD because WhatsApp failed to identify a legal basis for it. This finding highlights the golden rule of data protection under both the DPD and the GPDR: A firm cannot process any European resident's per-sonal data without a legal basis. To be compliant, privacy statements must identify one of the six previously listed legal bases for each reason given for processing personal data.

Consent 2.0. If WhatsApp were to rely on consent for its legal basis, ICO also ruled that it would be invalid, even under the DPD's more attainable consent standard. But the GDPR, the notion of valid consent shifts the para-digm significantly, such that the days of preselected boxes or opt-out consent are gone. According to the ICO's Draft Guidance on Consent, to qualify as a valid legal basis for processing under the GDPR, consent must be:

- *User friendly,* such that it is easy for data subjects to understand what is being asked of them and why.
- *Informed,* revealing to subjects what will be done with their data (i.e., the purpose).
- *Itemized/one consent per purpose,* which requires detailed and granular con-sent. Vague or sweeping descriptions of plans for personal data invalidate consent; each purpose also requires its own record of consent or other legal basis.
- *Unbundled,* such that the request for consent must be separate from any other information, rather than being integrated into a website's terms of use or privacy policy.
- *Opt-in,* so all consent is affirmative and explicit, rather than relying on pre-checked boxes.
- *Freely given,* meaning that consent is invalidated by a "take-it-or-leave-it" service offering.

Mind the Legal Basis. According to its privacy policy, in place at the time of the acquisition, WhatsApp reserved the "right to transfer or assign the information … collected from our users as part of [a] merger, acquisition, sale, or other change of control." Such a reservation of rights is a standard privacy policy provision. But it appears to have had no legal effect in this case. Therefore, marketers must recognize that they cannot necessarily rely on an exclusionary statement in their privacy policy to claim compliance with the GDPR.

References and Notes

1. Warren, S., & Brandeis, L. (1890). The Right to Privacy. *Harvard Law Review,* 4(5), 193–220. https://doi.org/10.2307/1321160. Available at http://groups. csail.mit.edu/mac/classes/6.805/articles/privacy/Privacy_brand_warr2.html.
2. According to Daniel Solove: "In the second latter half of the 19th century, newspapers were the most rapidly growing type of media. Circulation of newspapers rose about 1000% from 1850 and 1890, from 100 newspapers with 800,000 readers in 1850 to 900 papers with over 8 million readers by 1890." Solove, D. J. (2003). *The Origins and Growth of Information Privacy Law,* 10. Available at https://scholarship.law.gwu.edu/cgi/viewcontent.cgi?referer=https://www.google.com/&httpsredir=1&article=2091&context=faculty_publications.
3. Ibid., 11.
4. Warren, S., & Brandeis, L. at 1.
5. Ibid.
6. Ibid.
7. Pollak, M. (2018). A Short History of Wiretapping. *Nytimes.com.* Available at https://www.nytimes.com/2015/03/01/nyregion/a-short-history-of-wiretapping.html. Accessed May 23, 2018.
8. Kaplan, H. J., Matteo, J. A., Sillett, R., & Arkin Kaplan Rice, L. L. P. (2012). The History and Law of Wiretapping. In *ABA Sections of Litigation 2012 Section Annual Conference April* (pp. 18–20).
9. Olmstead v. United States. (n.d.). *Oyez.* Retrieved May 22, 2018, from https://www.oyez.org/cases/1900-1940/277us438.
10. Ibid.
11. Tokson, M. (2016). Knowledge and Fourth Amendment Privacy. *111 Northwest University Law Review, 139.* https://scholarlycommons.law.northwestern.edu/nulr/vol111/iss1/3.
12. Katz v. United States, 389 U.S. 347, 359 (1967).
13. Ibid., 360 (Harlan J., concurring).
14. Kerr, O. S. (2011). An Equilibrium-Adjustment Theory of the Fourth Amendment. *125 Harvard Law Review, 476,* 533–534.
15. McDonald, A. M., & Cranor, L. F. (2008). The Cost of Reading Privacy Policies. *ISJLP, 4*(543). Available at https://kb.osu.edu/bitstream/handle/1811/72839/ISJLP_V4N3_543.pdf?sequence=1&isAllowed=y.

16. See Reidenberg, J. R., Breaux, T., Cranor, L. F., French, B., Grannis, A., Graves, J. T., & Ramanath, R. (2015). *Disagreeable Privacy Policies: Mismatches Between Meaning and Users' Understanding.* Berkeley Tech. LJ, 30, 39; and Waldman, A. E. (2017). A Statistical Analysis of Privacy Policy Design. *Notre Dame Law Review Online*, 93, Forthcoming. Available at SSRN https://ssrn.com/abstract=2940505.

17. 15 USCS § 1681, *et seq.*

18. EPIC—The Fair Credit Reporting Act (FCRA) and the Privacy of Your Credit Report. (2018). Retrieved from https://epic.org/privacy/fcra/#introduction.

19. Department of Health, Education, and Welfare (HEW) Secretary's Advisory Committee on Automated Personal Data Systems (SACAPDS) (testimony of Kenneth A. McLean, May 18, 1972).

20. Federal Trade Commission Marketers of Criminal Background Screening Reports to Settle FTC Charges They Violated Fair Credit Reporting Act. (2013). Available at https://www.ftc.gov/news-events/press-releases/2013/01/marketers-criminal-background-screening-reportsto-settle-ftc.

21. Hoofnagle, C., & Honig, E. *EPIC Victoria's Secret GLBA Privacy Page.* Available at https://epic.org/privacy/glba/victoriassecret.html.

22. Office for Civil Rights, H. H. S. (2002). Standards for Privacy of Individually Identifiable Health Information. Final Rule. *Federal Register, 67*(157), 53181.

23. US Department of Health and Human Services. (2005). Security Standards for the Protection of Electronic Protected Health Information. Code of Federal Regulations.

24. Francke, G., Greene, A., & Williams, B. (2017, May 16). *Public Still Must Be Kept Private Under HIPAA.* Available at https://www.dwt.com/Public-Still-Must-be-Kept-Private-under-HIPAA-05-16-2017/.

25. Hoofnagle, C. J. (2016). *Federal Trade Commission Privacy Law and Policy* (p. 236). Cambridge University Press.

26. Monty Python—Spam. (2007). [Video]. Retrieved from https://www.youtube.com/watch?v=anwy2MPT5RE.

27. The Origin of SPAM (The Food) & Spam (The Email). (2009). Retrieved from http://mentalfloss.com/article/20997/origin-spam-food-spam-email.

28. Federal Trade Commission. *CAN-SPAM Act: A Compliance Guide for Business.* Retrieved from https://www.ftc.gov/tips-advice/business-center/guidance/can-spam-act-compliance-guide-business.

29. Manjoo, F. (2003). *The Day the Dinnertime Phone Calls Stopped.* Retrieved from https://www.salon.com/2003/07/15/do_not_call.

30. 47 U.S.C. § 227(b), (c).

31. Wireline Competition. (2018). *FCC Strengthens Consumer Protections Against Unwanted Calls and Texts.* Retrieved from https://www.fcc.gov/document/fcc-strengthens-consumer-protections-against-unwanted-calls-and-texts.

32. Mailing Standards of the United States Postal Service, Domestic Mail Manual (DMM). (2007). Available from https://pe.usps.com/cpim/ftp/manuals/dmm300/dmmtoc.pdf.

33. EPIC—Children's Online Privacy Protection Act (COPPA). Retrieved from https://epic.org/privacy/kids/.

34. Hoofnagle, C. J. (2016). *Federal Trade Commission Privacy Law and Policy* (p. 195). Cambridge University Press.

35. Ibid., 193.

36. Ibid.

37. Complying with COPPA: Frequently Asked Questions. (2018). Retrieved from https://www.ftc.gov/tips-advice/business-center/guidance/complying-coppa-frequently-asked-questions.

38. Security Breach Notification Laws. (2018). Retrieved from http://www.ncsl.org/research/telecommunications-and-information-technology/security-breach-notification-laws.aspx.

39. Lazzarotti, J. J., Gavejian, J. C., & Atrakchi, M. (2018, April 9). *State Data Breach Notification Laws: Overview of the Patchwork*. https://www.jacksonlewis.com/publication/state-data-breach-notification-laws-overview-patchwork.

40. See IBM Security, Ponemon Institute. (2017). *2017 Cost of Data Breach Study*. Available at https://www.ibm.com/security/data-breach. Accessed March 1, 2018

41. Ibid.

42. Bryan Cave. (2017). *2017 Data Breach Litigation Report* [Ebook] (1st ed., p. 6). Retrieved from https://d11m3yrngt251b.cloudfront.net/images/content/9/6/v2/96690/Bryan-Cave-Data-Breach-Litigation-Report-2017-edition.pdf.

43. Hoofnagle, C. J. (2016). *Federal Trade Commission Privacy Law and Policy* (p. 75). Cambridge University Press.

44. Hoofnagle, C. J. (2016). *Federal Trade Commission Privacy Law and Policy* (Chap. 1). Cambridge University Press.

45. 16 CFR para 2.31–2.32.

46. Federal Trade Commission. (2008). A Brief Overview of the Federal Trade Commission's Investigative and Law Enforcement Authority. Retrieved from https://www.ftc.gov/about-ftc/what-we-do/enforcement-authority.

47. Solove, D. J. (2008). *Understanding Privacy* (p. 613). Harvard University Press.

48. Ibid., 121.

49. Ibid., 620.

50. E.g., In re Eli Lilly & Co., 133 F.T.C. 763 (2002) (complaint) (charging company with breaking privacy agreement by disclosing customers' personal information).

51. E.g., In re HTC Am. Inc., FTC File No. 122 3049, No. C-4406 (F.T.C. June 25, 2013). Available at http://www.ftc.gov/sites/default/files/documents/cases/2013/07/130702htccmpt.pdf (charging company with failing to mitigate security vulnerabilities when providing third parties with sensitive information); In re Microsoft Corp., 134 F.T.C. 709, 715 (2002) (complaint) (charging

company with collecting information beyond that provided for in privacy policy).

52. E.g., In re Genica Corp., FTC File No. 082 3113, No. C-4252 (F.T.C. March 16, 2009). Available at http://www.ftc.gov/sites/default/files/documents/cases/2009/03/090320genicacmpt.pdf; Microsoft, 134 F.T.C. at 711–712.

53. E.g., In re Compete, Inc., FTC File No. 102 3155, No. C-4384 (F.T.C. February 20, 2013). Available at http://www.ftc.gov/sites/default/files/documents/cases/2013/02/130222competecmpt.pdf (charging company with failing to strip personal information before transmission of data to servers).

54. In re Toysmart.com, FTC File No. X00 0075, No. 00-11341 RGS (F.T.C. July 21, 2000) (Swindle, Comm'r, dissenting). Available at http://www.ftc.gov/sites/default/files/documents/cases/toysmartswindlestatement_0.htm.

55. Federal Trade Commission. In the Matter of Uber Technologies, Inc., Complaint. FTC File No. 152 5034. Retrieved from https://www.ftc.gov/system/files/documents/cases/170206_vizio_stipulated_proposed_order.pdf.

56. Hill, K. (2014). 'God View': Uber Allegedly Stalked Users For Party-Goers' Viewing Pleasure (Updated). Retrieved from https://www.forbes.com/sites/kashmirhill/2014/10/03/god-view-uber-allegedly-stalked-users-for-party-goers-viewing-pleasure/#4cd117ed3141; http://www.cosmopolitan.com/lifestyle/a8495499/uber-using-god-view-tool-to-spy-on-celebs/.

57. Sims, P. (2014). Can We Trust Uber?—Silicon Guild. Retrieved from https://thoughts.siliconguild.com/can-we-trust-uber-c0e793deda36.

58. Federal Trade Commission. Analysis of Proposed Consent Order to Aid Public Comment In the Matter of Craig Brittain File No. 132 3120. Federal Trade Commission. Retrieved from https://www.ftc.gov/system/files/documents/cases/150129craigbrittainanalysis.pdf.

59. E.g., In re H&R Block, Inc., 80 F.T.C. 304, 304–09 (1972) (complaint) (discussing notice-related reasons for FTC violation). E.g., In re Sears Holdings Mgmt. Corp., FTC File No. 082 3099, No. C-4264 (F.T.C. Aug. 31, 2009) [hereinafter Sears Complaint]. Available at http://www.ftcgov/sites/default/files/documents/cases/2009/09/090604searscmpt.pdf.

60. Federal Trade Commission. In the Matter of Lenovo (United States) Inc., Complaint. FTC File No. 152 3134. Retrieved from https://f.datasrvr.com/fr1/018/84049/1523134_lenovo_united_states_complaint.pdf.

61. Hoofnagle, C. J. (2016). Federal Trade Commission Privacy Law and Policy (p. 160). Cambridge University Press.

62. (15 U.S.C. Sec. 45(n)).

63. Federal Trade Commission. In the Matter of Facebook, Inc., (United States) Inc., Complaint., FTC File No. 092 3184 (July 27, 2012). Retrieved from https://www.ftc.gov/sites/default/files/documents/cases/2012/08/120810facebookcmpt.pdf.

64. Federal Trade Commission. In the Matter of Vizio, Inc., Stipulated Order for Permanent Injunction and Monetary Judgement. Retrieved from https://www.

ftc.gov/system/files/documents/cases/170206_vizio_stipulated_proposed_order. pdf.

65. Francke, G. (2018). *UK Facebook Investigation Offers Early GDPR Lessons— Law360*.   Retrieved   from   https://www.law360.com/articles/1031676/ uk-facebook-investigation-offers-early-gdpr-lessons.

66. Whatsapp Blog. https://blog.whatsapp.com/499/Facebook? February 19, 2014.

Part II

Defensive Strategies: Protecting from Threats

4

Big Data's Marketing Applications and Customer Privacy

Introduction

By the time you read this chapter, the odds are that some of the content and examples will be outdated. Although we wrote it using the most innovative insights of the time, trends in data and analytics are evolving at such a rapid pace that nearly every technology that was just recently groundbreaking quickly becomes obsolete. Did your firm embrace machine learning? Turns out you need deep learning. Investing in virtual reality? The space to be is actually augmented reality. Think you've harnessed your customers' clickstream and browsing behavior? The hot new data are their unstructured voice text and video streams [1].

Although these technologies and applications continue to advance at breakneck speed, we argue that the marketing questions and objectives that underpin data analysis actually remain relatively stable. That is, most marketing organizations employ big data and analytics to understand their customers, enhance the customer experience, and ultimately form long-term, mutually beneficial relationships with customers. These ideas represent marketing's bread-and-butter objectives, since the earliest days of loyalty punch cards. According to a 2014 Forrester Research Report [2], the true potential of big data technologies is the ability they offer to help firms achieve traditional marketing goals through enhanced connections to their increasingly fragmented customer bases. Increasingly, business experts advocate that novel technologies should be less about the data, and more about how data complements real business objectives. For example, data strategist Bernard Marr notes,

© The Author(s) 2019

R. W. Palmatier and K. D. Martin, *The Intelligent Marketer's Guide to Data Privacy*,
https://doi.org/10.1007/978-3-030-03724-6_4

"No company should start a big data project right now. But, let me be clear. It's not because I predict a major downturn in the economy or an incredible innovation that will change the way a data project should be run. It's really simple. If you plan a big data project, you're missing the most important aspect. It shouldn't be about the data; it should be about your business" [3].

Companies recognize the value of data-driven marketing applications for achieving customer-focused objectives. Spending on big data and analytics is on track to exceed $200 billion by 2020 [4], and 80% of recently surveyed companies report that their data-focused initiatives have been successful. Nearly half of the executives that responded to this survey attest to tangible, quantifiable performance benefits, and only 1.6% of the respondents reported that their big data initiatives were not successful [5]. As data become ever more abundant, firms also indicate the heightened sophistication of their analytical capabilities to make sense of the information. Big data-enabled artificial intelligence (AI) applications in particular have prompted a greater than 10% boost in customer satisfaction for three-quarters of the firms that employ these techniques [6]. As a result, 84% of marketing organizations or divisions surveyed reported either implementing or expanding their machine learning and AI capabilities in 2018.

Evidence in support of big data's marketing benefits continues to build, creating pressures for companies to infuse their own practices with these technologies. Far less frequently studied are the inherent trade-offs that these practices create for both the firm and its customers. As we will detail in this chapter, big data and analytics sometimes create situations in which customers and marketers are directly at odds, but in other cases, they enable customers and marketers to create shared value through collaboration. We explore these tensions and trade-offs through the explicit lens of marketers' customer data use, that is, not the specifications of the technologies or the data themselves. Despite the many mutual benefits that big data create, what firms do with these data has the potential to create substantial and unforeseen risks for their customers. In turn, customer risks simultaneously create important opportunities for firms to take the lead on privacy, minimize customer vulnerability, and gain competitive advantages. As one data analytics expert explains, "Organizations that are flexible enough to manage and transform the data into useful business intelligence, this represents a significant opportunity to gain a competitive advantage (or remain competitive)" [7]. To this point we add an important qualifier: Companies must do so in a way that minimizes customer vulnerability.

This chapter emphasizes the marketing applications and uses of customer data. We begin with a basic explanation of the predominant modes by which marketers entice customers to exchange their information for some form of value. This section also breaks down the various marketing benefits realized by specific big data applications, while highlighting possible customer risks. Next, we carefully explore the inherent tensions between marketing's big data and analytics applications and customer privacy. Many of the unintended consequences of marketing analytics for customer privacy and vulnerability likely are contradictory with the marketing organization's stated, more benign goals, so an explicit understanding is critical. Finally, we examine how current AI applications in marketing offer deeper customer insights through powerful technologies but also escalate the potential for customer vulnerability in dangerous ways. Our hope is that with this awareness, marketers can avoid privacy pitfalls, analytics program failures, and the potential for more egregious data privacy crises. We address big data customer–firm trade-offs in this chapter; Chapter 5 offers concrete solutions for protecting customers from vulnerability and simultaneously inoculating the firm against related harms.

> Many of the unintended consequences of marketing analytics for customer privacy and vulnerability likely are contradictory with the marketing organization's stated, more benign goals, so an explicit understanding is critical.

Advances in Marketing Applications

As big data and analytics increasingly become part of firms' marketing strategies, it is apparent that these applications offer benefits in relation to various marketing activities. In this section, we explain how data-driven insights transform marketing strategy and customer relationships. We begin with discussion of big data applications in the vast marketing environment, reviewing some of the societal and customer forces they may inform. We transition to understanding how data-enabled technologies facilitate precision target marketing and conclude by discussing their ability to enhance the full portfolio of marketing tactics available to firms.

Understanding the Marketing Environment

The environment in which marketers operate is inherently broad and diverse. Depending on their business scope, marketers might be responsible for navigating a variety of external concerns and pressures. Environmental forces include natural environmental concerns and sustainability, the regulatory and political environment, technological factors and access, as well as economic and sociocultural forces. Using big data and analytics to understand the marketing environment can benefit the firm's operations and practices directly. For example, analytics that allow a company to design a more sustainable supply chain or attract a more diverse workforce can offer immediate benefits to the company, such as operational efficiencies and improved morale. These examples illustrate the important environment-facing benefits that can benefit customers indirectly, because the company issues better product offerings and increased value.

When customers interact directly with firm-related environmental factors, they likely share their opinions about controversial or polarizing issues. With data and analytics, marketers can determine how their customers feel about a broad set of issues and choose to align themselves accordingly. For example, academic research shows that consumers increasingly embrace their political ideology as a focal part of their identity [8] and make purchase decisions on the basis of those beliefs, by considering company values and how they perceive them to align with their personal values [9]. Likewise, companies' stances on a variety of social issues—from marriage equality to sweatshop labor to racial biases—create customer motivations for both boycotts (i.e., avoiding companies' products or services) and buycotts (i.e., purposely purchasing their products or services). These influential consumer behaviors require a clear understanding of the marketing environment, because firms risk offending customers when they embrace or ignore certain social issues [10]. Big data and analytics are key means by which firms can navigate this treacherous terrain, especially considering how much personal information about these sensitive matters is readily accessible through social media analytics and information that customers freely provide [11]. Being aware of social issue landmines is of critical importance to modern companies; the best practices for taking a principled stand on social issues are still evolving though.

Precision Targeting

Marketers have learned that the available troves of personal data can help them understand and navigate sensitive situations by revealing customers' beliefs and ideologies. These data also enable them to segment specific markets and target desirable customer groups with great precision. The ability to understand customers, beyond simply defining their rudimentary characteristics such as age, gender, and geography, means that marketers can deploy their resources more effectively and reach customers more strategically. In terms of segmentation, marketers can formulate internal definitions of customer groups along meaningful breakpoints; relevant niche groups also can be understood and reached in ways that were not possible before. This understanding benefits precision target marketing and informs the product design process too.

Such a precise understanding of customers and the ability to target them with appealing products and services in turn is a key promise of big data. Accurate targeting is a critical step toward avoiding marketing inefficiencies and reaching the people with whom the firm can enter into mutually beneficial exchanges. However, as we describe in the next section, highly targeted marketing also can create skepticism and or a sense of creepiness; customers come to think that the marketer knows them a little too well. Therefore, marketers are responsible to limit their activities to reaching customers who are likely to value the specific offering, rather than engaging in targeting on the basis of personal or trivial attributes—simply because they can. The key question is whether the identified differences among customers are meaningful differentiators, given the risks that specific personalization create. If they are, they may be worth targeting. Such an exercise can follow a pattern similar to that used to evaluate the features and attributes of a new product. To return to our example from Chapter 1 about OfficeMax's mistake in targeting the grieving father, if the firm had taken the care to ask whether the knowledge was a meaningful customer differentiator, given the risk, we probably would not be talking about it today.

Big Data's Role in Value Creation, Communication, and Delivery

Data analytic insights can enhance marketing executions, because they help marketers undertake more effective value creation through improved offerings and customer experiences. Technology-enabled platforms allow customers to co-create products and services, embedding novel design specifications into the very fabric of the firm's innovation. These analytics in turn can be monitored to support real-time modifications, reflecting consumers' interactions with various features and attributes. Companies ranging from Ford Motor to Nike to Julep Cosmetics [12, 13] use big data and real-time customer responses to make critical product improvements. Similarly, marketers can test pricing strategies, options, and price points using data analytics.

Dynamic pricing applications even allow marketers to change prices in real time, in accordance with customer demand and other external factors. This very logic underpins Uber's surge pricing mechanism, which adjusts fees to reflect peak demand times and destinations [14], as well as airlines' ability to charge different customers different fares for identical flights [15].

Value communication to customers often involves targeted advertising, which represents another important application of customer insights gained from big data. The ability to reach customers through the media they use, describing products and services ideally tailored to them, represents an ideal form of big data marketing. Yet similar to other applications of big data, targeted advertising is developing its own reputation for creepiness, by suggesting too intimate a level of familiarity with a customer, or else providing humorously faulty "You might also like…" recommendations.

New approaches to convey highly targeted ads appear to be on the horizon. For example, in recent research that addresses consumer backlash to targeted ads, researchers establish that companies need to be more transparent in their uses, grant customers control over the personal settings that lead to a suggestion for a particular ad, and provide a clear justification for using this method of communication [16]. Alternative but still closely targeted approaches to communicating with customers include geofencing, which can alert customers to promotions in physical locations in their geographic proximity in real time, such that it combines the benefits of big data for both value communication and value delivery. Geoconquesting takes this application a step further and targets communications at customers when they approach competitor locations in real time [17].

Value delivery using big data accordingly has evolved beyond just getting products and services to customers in the desired form and at a time most convenient to them, though this simple premise still informs Amazon's home delivery options. Customers can rely on options that grant delivery drivers for Amazon special access to the front doors of their homes or the trunks of their vehicles. Multiple applications also give consumers improved access to relevant art and culture offerings. Using music preferences gleaned through services like Spotify or Apple Music or even live streams of National Public Radio, marketers can send consumers personalized information about upcoming performances and events, help them purchase tickets in a few clicks on a smartphone, and complete the payment using a preloaded form through Apple Pay or Google Wallet. The ticket, downloaded through an app, makes customers' access to a favorite musician as effortless as a few short clicks. Then before and during the show, customers might access the musician's personalized app and engage with the artist to an even greater extent, such as previewing the concert's playlist or sharing photos and other content via social media. The next time this artist comes back to town, reconnecting and beginning the cycle again would be no more complicated than an informative email, text message, or social media alert.

In this simple example, an idealized version of the marketing strategy planning process unfolds, showcasing the multiple ways big data enable a connected series of marketing applications. Yet as described throughout this book, there are inherent customer privacy trade-offs embedded in each of these steps and applications. We address those tensions in the next section, exploring the benefits and challenges created for not only customers but also the firms and marketers that rely on big data insights.

Big Data's Inherent Tensions

Our previous review samples just a few of the myriad marketing applications enabled by big data. Already these applications have evoked positive marketing outcomes and provided evidence that firms should devote resources to them. Data and analytics can offer marketers a clear sense of their progress toward important marketing goals and objectives. Yet as Bruno Aziza, a big data technology expert, argues, "Big Data will be less about technology. It will be more about management practices and processes [18]." We thus look beyond marketing applications to consider the benefits and challenges for companies created by using big data in marketing. We perform a parallel analysis of the benefits and challenges for customers, created by firms' uses of big data. Both feature in Exhibit 4.1, specified according to some of the most recent advances in targeting technologies.

Company Benefits

Technologies for analyzing big data offer marketers many advantages in terms of knowing their customers, in ways that were not possible until recently. A massive leap involves compiling customer data and personal information from various sources. Using their own repositories, as well as acquisitions through third-party data providers, marketers have access to a broad portfolio of customer information. Data and information about marketing initiatives, activities, and spending, coupled with firm sales and point-of-sales (POS) data, help marketers understand what works and what does not. When this information is combined further with information

Exhibit 4.1   Big data benefits and challenges for customers and companies

Company objective	Customer benefit or challenge?	Critical takeaway
Knowing customers with accuracy	Can bring customer benefits and risks	Companies need to use caution before exploiting customer information to add value or enhance their experience. Greater accuracy in customer interactions can create stronger relationships and engagement, but companies run the risk of appearing creepy.
Pooling customer data from various sources	Can bring customer benefits and risks	Companies must be transparent about which customer data they have access to and how they are using them. Customers appreciate disclosures about why companies use particular information. It can be particularly helpful when offering customers benefits that reflect their prior behavior.
Ease of experimentation and forum for customer co-creation	Co-creation brings mostly customer benefits	Platforms for customer co-creation offer mutual benefits. These new modes of interaction can create innovations that exceed customer expectations and foster delight. Co-creation models can span geographies and help solve important global problems.
Technologies that remove human error from business practices	Improved company accuracy and efficiency should benefit customers	Greater operational accuracy and efficiencies provide better customer experiences. These efficiencies and reduced waste allow the firm to pass on potential cost savings, for added customer benefits. Positive experiences likely prompt further engagement and loyalty.
Identifying, migrating, or removing unprofitable customers	Bring customer risks for poor service and alienation	Technologies that allow companies to identify unprofitable customers create the potential for abuse. Customers may receive poorer service experiences and terms. Fired customers may become alienated from the broader marketplace for such services.
Customer data as unanticipated and low-cost profit source	Customer lacks control of her/his information, creating risk	Customers currently lack the ability to be fairly compensated for uses of their personal information as profit sources. Companies continue to reap these benefits, but this profit source is risky, especially in light of proposed legislation and demands for customer payment (see Chapter 1).

derived from social media chatter, stock market performance, weather conditions, the economy, and global events, for example, marketers can make accurate decisions based on a more complete picture of reality.

Beyond data and the corresponding tools they offer for understanding customers, these analytical techniques create other marketing benefits too.

Real-time feedback and co-creation enabled through big data can be important boons to innovation. When paired with the ability to micro-target specific and unique customer needs, they allow marketers to identify increasingly fine-grained but nonetheless differentiable and profitable customer groups. Beyond creating new products and services, with successful micro-targeting a firm can disrupt entire industries. This type of innovation is already prevalent in pharmaceutical sectors [19], where rare drugs command huge profits [20]. Companies such as Gilead, Vertex, and Alexion have identified rare disorders using precision marketing technologies to reap substantial profits, even though the target market of patients suffering from rare conditions is small in number. Yet their needs are specific, so an effective solution, targeted directly at them, has incomparable value. For example, Soliris developed a medication to treat a rare blood disease that costs patients more than $400,000 per year. Yet because it is the only drug of its kind, even the small number of prescriptions make this a highly profitable product for the firm [21].

The analytical models used to obtain important marketing insights also can be modified in response to new data or parameters, in real time. This real-time evaluation and monitoring encourages an experimental mindset, reducing the firm costs and risks traditionally associated with strategic experimentation. Moreover, at the top levels of the firm, the breadth and depth of data evaluated using sophisticated analytics establishes greater marketing accountability. Using a variety of information sources, marketers can demonstrate their effectiveness with various customers and marketing initiatives. This critical benefit gives marketers the ability to make an evidence-based case to support future, increasingly intentional resource allocations [22].

Company Challenges

A significant challenge associated with collecting customer data is identifying the most fruitful and informative data to gather, analyze, and interpret. Marketers tout big data's insights; we hear far less about the "useless" data taking up valuable cloud or other storage space, waiting to be analyzed. Far too often though, companies tend to collect vast volumes of data, even without having any certainty about how to apply them—or even knowing if they are applicable [23]. Modern corporations have amassed so much data that traditional methods that rely on human power simply cannot keep pace; advanced analytical techniques are needed to make sense of them. It is precisely this problem that AI aims to solve, though it introduces its own unique concerns, as we describe later in this chapter.

Perhaps even more risky is firms' increasing accumulation of customer data as a profit source. We noted this point briefly in the book's introduction, but we reemphasize it here because of its notable threat to both firms and consumers. The complex legal and political landscape (Chapter 3), coupled with widespread disdain for firms' data monetization (Chapter 2), means that relying on such a business model is fraught. Facebook and Amazon may have embedded data access and use into their very fabric and business model, and customers appear to accept this premise when the exchange parameters are clear. But these are standouts; when companies attempt to capitalize on the serendipitous profitability of data they already had been collecting by selling them to third parties, customers tend to balk [24]. For example, automakers have long possessed volumes of driving data that advertisers, insurers, and others are eager to access. According to one company hoping to connect marketers to automakers and their data, drivers of mid-to low-tier cars are highly receptive to in-vehicle ads in exchange for free services [25]. In the past, this third-party profiteering occurred mostly behind the scenes, but awareness of data monetization is becoming far more mainstream. As McKinsey notes, it represents an important and growing company revenue source [26]. The evidence of customer backlash and the increasing potential for regulation suggests though that relying on the

> The evidence of customer backlash and the increasing potential for regulation suggests though that relying on the commoditization and monetization of data as a focal revenue source will lead to an unhealthy dependence and heightened performance risk.

commoditization and monetization of data as a focal revenue source will lead to an unhealthy dependence and heightened performance risk.

Finally, despite the profits, applications, and analytical power that data companies are amassing, many organizations still have insufficient human resources and skills to make sense of their own data. Analytics experts have hopefully declared the *data scientist* as the "sexiest" profession of our time [27], yet *data translators*, or people who can understand the output of data science tools and convey its important takeaways to key functional units and decision makers remain in great demand and insufficient supply. Investments in data science and analytics programs by universities, trade schools, and firm-internal technical departments have created a glut of expertise related to processing and analyzing data [28], not translating those data into real-world implications. Considering the early potential already demonstrated by machine learning, data scientists also suggest the potential for growing reliance on data-enabled technologies. Still, many companies continue to lack data translators who can develop sophisticated analytical insights, then use and share them, in a language understood by laypeople, managers, users, and customers.

Customer Benefits

The company–customer tensions we describe are not mutually exclusive; some company benefits and challenges also appear as customer benefits and challenges. First, customers' ability to co-create value in tandem with the firm, using technologies enabled by big data, has enormous potential for achieving product customization and superior value offerings. The benefits to both customers and companies from partnering in product and service development efforts, using data-enabled applications in real time, cannot be overstated. Experts already have imagined solving significant global problems, from poverty to health care to climate change, through such interfaces. The MIT Media Lab Director Alex Sandy Pentland touts big data's ability to transform the world, its social institutions, and all modes of interaction [29]. The potential for problem solving through co-creation stems from big data's beneficial ability to recognize previously complex problems or invisible consumer groups that have wants and needs that marketers can meet. These capabilities are widely touted in pharmaceutical sectors, as we noted previously, but micro-targeting also might benefit disease prevention efforts, financial well-being, microlending, nutrition, athletics, and human resource development, among many others. For example, micro-targeting in support of microlending initiatives could specify precisely which households in impoverished areas are most likely to repay their loans, as well as which ones are most likely to leverage the funding in productive ways. Micro-targeting in areas without access to healthy options, such as food deserts, might identify profitable mobile technologies for bringing better foods to receptive households. Finally, micro-targeting may be the technology best poised to identify high potential athletes in even the most unlikely parts of the world.

Second, the potential for misuse is inherent to consumer profiling, yet extracting personal information from novel sources also may allow companies to move beyond traditional demographic or other profile characteristics to offer value in ways that are more consistent with consumers' actual preferences or behavior. For example, in car insurance markets, premiums could be linked to actual driving performance through smart vehicles, so that people get charged according to how they actually operate the car. Utility companies might tailor their efficiency programs to customers, depending on how attentive they are to their energy usage, as gauged by smart home devices such as the Nest. Customers who are truly loyal, ardent users of a firm's products and services also might accumulate rewards that set them apart from the company's traditional user base, which should deepen their engagement. These examples showcase ways in which analytical techniques offer customers superior benefits and experience.

Moreover, data and analytics can improve the firm's behind-the-scenes performance, which should enhance customer experiences. AI and machine learning largely remove human error from the equation, which is particularly valuable in medical settings and automotive service markets. Smart city technologies support infrastructure monitoring, so that city engineers can adjust lanes and lights to match traffic flows in real time and use predictive analytics to anticipate disruptions. The benefits of a more efficient morning commute can go a long way to improving people's personal well-being and workplace productivity. Even as the customer value created by data-enabled technologies continues to surge, the benefits come at a cost; in particular, most applications require customers to relinquish some personal information, much of which is sensitive, as the key trade-off in these exchanges.

Customer Challenges

Such widespread acquisition of personal information creates customer vulnerability, simply due to data access [30]. Marketers increasingly use customer data for new applications, to reach customers in novel, unexpected ways; as they do so, customer vulnerability only increases. The continuum we introduced in Chapter 2 represents the foundation for understanding these customer challenges and risks. Knowing their data is "out there" but not knowing how they will be used makes consumers feel powerless and worried about the potential for harm. Even with these pressing worries, the sheer volume of personal information that exists beyond each person's immediate access means no one can maintain complete personal control. Managing the rights to personal information across business, government, medical, educational, and other spheres just is not possible. An unintended consequence of the overwhelming volume of personal data is the human complacency it creates. An inability to manage all data-extracting sources and uses creates a sense of futility, which can lead customers to cease any personal data stewardship. When this occurs, customers become increasingly likely to suffer, as unknowing targets of inappropriate or fraudulent marketing communications, without proper defenses to avoid price discrimination, detect cross-selling, or differentiate genuinely valuable offerings from those that only seem ideally customized.

Marketing may appear tailored but fail to deliver meaningful value; micro-targeting instead can create a reverse effect. It allows marketers to understand their customers' needs and preferences with extreme precision and differentiation, so it can get desired products and services into niche customers' hands, but only at the salient risk of seeming too intimate or even creepy.

> Micro-targeting instead can create a reverse effect. It allows marketers to understand their customers' needs and preferences with extreme precision and differentiation, so it can get desired products and services into niche customers' hands, but only at the salient risk of seeming too intimate or even creepy.

Customers appreciate customized products and services, but marketers can quickly go too far in their efforts to personalize unique customer offerings [16].

Micro-targeting also can create big picture customer or societal harms by reinforcing certain preferences, ideas, and beliefs in overly narrow ways. For example, due to big data, people today consume news media in ways most conducive to their core beliefs, such that they receive articles and content that align with their personal worldviews. The failure to offer different or contrary perspectives gives people limited insight into the full complement of news and information. This self-reinforcing cycle risks the proliferation of potentially dangerous ideologies and radical mindsets, as well as deepening political and ideological divides [31].

Finally, big data can exclude undesirable or low performing customers from receiving the firm's marketing efforts, based on personal information such as income and other demographics. In the same way that loyalty programs and customer relationship management technologies enable relationship formation and retention of the company's "best" customers, these programs can use data to shut out the firm's "worst" customers, potentially in damaging or unethical ways. Similarly, RFID and smart chip technology support the ability for retailers to identify a person's location in the store immediately and offer prices or services in a manner that may discourage their future patronage, just as modern technologies encourage such patronage by desirable customers [32]. Advice for marketers about how to manage their worst customers is far less common than advice for promoting relationships with good customers [33]. Firing bad customers might become more precise with accurate data that allows firms to identify customers who—regardless of their actual need for the products that the firm provides—offer little to no payoff for the firm.

For example, similar technologies allow financial institutions and car dealerships to charge "risky" customers higher interest rates and offer suboptimal products, services, and negotiated terms [34]. An already popular application of this thinking comes from the third-party data firm Retail Equation, which calculates customer risk scores on the basis of various aspects of a person's product return history [35]. Retailers such as Best Buy, CVS Health, Home Depot, and J. C. Penney use Retail Equation's services to implement policies for recouping return losses, such as limiting future return flexibility, requiring receipt verification, creating hard no-return time periods, and even charging higher prices to risky customers in hopes of discouraging repeat offenders. Far more discriminatory and potentially harmful exclusionary marketing practices have become a salient flipside to the highly successful and profitable loyalty programs firms implement. Firing customers on the basis of big data may be the next wave for cost reduction efforts, and machine learning will continue to enhance the sophistication, precision, and automation of these practices. Once the systems have been put into place, managers put their faith in their reliability, such that reevaluations or re-inclusion of fired customers becomes increasingly unlikely.

Potential and Peril of Artificial Intelligence in Marketing

Novel Applications

Data analytics experts such as McKinsey Global [28] predict that emerging applications and implementation practices enabled through AI will provide firms with substantial competitive advantages and even prove to be industry disruptors. Some of the predicted benefits resemble traditional value exchange models, such as the sense that big data will enhance discovery and innovation, and others imply substantial customer vulnerability risks. The level of customer information required to create radical personalization implies that firms will gain or already have access to vast amounts of highly sensitive personal details. The predicted real-time matching and massive data integration initiatives also suggest that firms will need to monitor their customers' behavior constantly and connect people's lives and current status with diverse data from a variety of sources. The risks of abuse of this type of customer monitoring already have been manifested tragically, such as in reports of domestic violence enabled by smart home technologies [36]. Such powerful uses of data-enabled technologies mean that companies are poised to acquire an unprecedented level of information about customers and exploit those insights for a level of marketing personalization unlike any we have even seen or imagined to date. Careful consideration of unintended consequences is less evident. We make a first attempt at doing so in Exhibit 4.2, noting many of the vulnerability risks posed by novel data-enabled technologies.

To be fair, in a recent Forrester Research Report on the Global State of Artificial Intelligence, of 598 professionals surveyed, the top AI applications

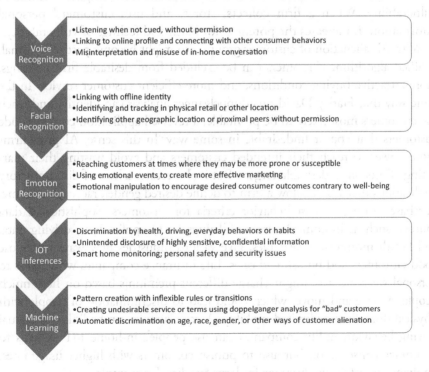

Exhibit 4.2 Customer vulnerability escalation through artificial intelligence

included some noble marketing purposes. The primary benefit sought was improving the customer experience (57%), followed by improving existing products and services (44%). Increasing value extracted from customers makes the list, but at a lower 20% [37]. Companies that currently employ AI applications for marketing also are realizing profound effects. For example, KLM Airlines transformed its customer service program using voice-enabled technologies that respond quickly and accurately to customer complaints. Harley-Davidson uses AI tools for doppelganger analyses of current high-value customers to attract new, similar, high-value customers, producing a nearly 3000% increase in sales in the first months of its use [38].

Customer Vulnerability Escalation

As success stories such as Harley-Davidson's continue to occupy the business press, AI applications are likely to spread to a wider variety of firms. Practically speaking, when firms accumulate vast data volumes, AI applications may be the best suited methods for managing the sheer size and scope of these data. Yet their expanded uses simultaneously heighten customer vulnerability. When a firm collects, stores, and uses customers' personal information, it increases the potential for harm and feelings of vulnerability.

With AI, alienation of customer groups also might become part of a formal, codified algorithm. Customers can be excluded from desirable firm offerings, more attractive buying conditions, and more efficient customer service. In the same way that Harley-Davidson's doppelganger analysis can identify and reach the customers most likely to be profitable, a reversed application could exclude customers that appear undesirable in some way. In this sense, AI gives firms more power to reach their intended customers and avoid wasting their marketing efforts on undesirable targets. But it also creates a host of discriminatory applications that would allow a firm to define desired gender, race, age, income, purchase history, or past behavior criteria for customers. Sophisticated data sources, such as in-home Internet-of-Things (IoT) applications, driving data, and health monitoring wearables, give various industries new ways to choose good customers and punish bad ones. Life insurance companies with access to personal wearable data might charge different premiums based on how much people exercise and move, which could lead to discrimination of people with physical disabilities or limited mobility. Auto insurance can be linked to actual driving behavior; utility companies can use people's in-home IoT systems to encourage conservation but also to punish customers with higher usage rates, implying an unfair disadvantage for large families, for example.

AI gives firms more power to reach their intended customers and avoid wasting their marketing efforts on undesirable targets. But it also creates a host of discriminatory applications that would allow a firm to define desired gender, race, age, income, purchase history, or past behavior criteria for customers.

Many applications enhance firms' efficient marketing efforts, but the unintended consequences of punishing "undesirable" customers can leave those who already are most vulnerable increasingly at risk. What is perhaps even more dangerous is that, unlike the felt vulnerability that customers experience when they know about firms' data access and use, with AI the customer may never know that she or he has been excluded from better product and service offerings. A customer might simply accept higher prices, longer queues, and unmanageable wait times as a personal reality, with no idea that they are having a subpar experience simply due to where they map on the firm's algorithms. If they do realize their low status, it is difficult to imagine how the firm's well-established algorithms might accommodate those customers back into the good graces of desirable firm offerings.

We refer to this risk as customer vulnerability escalation, and we argue that AI has the potential to create extreme customer vulnerability. Alienation is but one method firms might use to punish and exclude undesirable customers. Many other unintended consequences of AI that heighten vulnerability exist, in reality or in theory. When intelligent marketers understand these practices and their potential for misuse, they can better prevent the emergence of unintended consequences.

In the chapter that follows, we thus offer strategies to help firms defend against customer vulnerability of all types. We illustrate how building on the transparency and control framework introduced previously can establish a strong defensive approach.

References

1. Karr, D. (2017, March 8). What Is Big Data? What Are the Benefits of Big Data? *MarTech*. Available at https://martech.zone/benefits-of-big-data/. Accessed June 12, 2018.
2. Forrester Consulting. (2014). *Marketing's Big Leap Forward: Overcome the Urgent Challenge to Improve Customer Experience and Marketing Performance.* Available at www.forrester.com. Accessed June 13, 2018.

3. Marr, B. (2017, August 17). Why No Company Should Start a Big Data Project Right Now. *Forbes*. Available at https://www.forbes.com/sites/bernard-marr/2017/08/17/why-no-company-should-start-a-big-data-project-right-now/#59604a8962d6. Accessed June 13, 2018.

4. Press, G. (2017, January 20). 6 Predictions for the $203 Billion Big Data Analytics Market. *Forbes*. Available at https://www.forbes.com/sites/gil-press/2017/01/20/6-predictions-for-the-203-billion-big-data-analytics-market/2/. Accessed June 12, 2018.

5. NewVantage Partners. (2017). *Big Data Executive Survey 2017*. Available at http://newvantage.com/wp-content/uploads/2017/01/Big-Data-Executive-Survey-2017-Executive-Summary.pdf. Accessed June 12, 2018.

6. Columbus, L. (2018, February 25). 10 Ways Machine Learning Is Revolutionizing Marketing. *Forbes*. Available at https://www.forbes.com/sites/louiscolumbus/2018/02/25/10-ways-machine-learning-is-revolutionizing-marketing/#76215f645bb6. Accessed June 12, 2018.

7. Foote, K. D. (2017, November 28). Big Data Trends for 2018. *DataVersity*. Available at http://www.dataversity.net/big-data-trends-2018/. Accessed June 12, 2018.

8. Khan, R., Kanishka, M., & Vishal, S. (2013). Ideology and Brand Consumption. *Psychological Science, 24*(3), 326–333.

9. Ordabayyeva, N., & Fernandes, D. (2018). Better or Different? How Political Ideology Shapes Preferences for Differentiation in the Social Hierarchy. *Journal of Consumer Research, 45*(2), 227–250. https://doi.org/10.1093/jcr/ucy004.

10. Smith, N. C., & Korschun, D. (2018). Finding the Middle Ground in a Politically Polarized World. *MIT Sloan Management Review*. Available at https://sloanreview.mit.edu/article/finding-the-middle-ground-in-a-politically-polarized-world/?utm_source=twitter&utm_medium=social&utm_campaign=sm-direct. Accessed June 14, 2018.

11. Merrill, J. B. (2016). Liberal, Moderate or Conservative: See How Facebook Labels You. *New York Times*. Available at https://www.nytimes.com/2016/08/24/us/politics/facebook-ads-politics.html. Accessed June 14, 2018.

12. Post, R. (2014). Ford and Nike Use Big Data to Make Smarter Sustainable Design. *The Guardian*. Available at https://www.theguardian.com/sustainable-business/ford-nike-big-data-smart-sustainable-design. Accessed June 15, 2018.

13. Underwood, R. (2014). Putting Data in Design. *Inc*. Available at https://www.inc.com/magazine/201312/ryan-underwood/internet-companies-using-data-for-design.html. Accessed June 15, 2018.

14. Kedmey, D. (2014). This Is How Uber's 'Surge Pricing' Works. *Time*. Available at http://time.com/3633469/uber-surge-pricing/. Accessed June 14, 2018.

15. Rizzo, C. (2018). Airlines Want to Start Charging Customers Based on Who They Are—And It Means Everyone Could Be Paying Drastically Different Prices. *Business Insider*. Available at http://www.businessinsider.com/

airlines-charging-different-fares-for-different-people-2018-2. Accessed June 14, 2018.

16. John, L. K., Kim, T., & Barasz, K. (2018, January–February). Ads That Don't Overstep. *Harvard Business Review.* Available at https://hbr.org/2018/01/ads-that-dont-overstep. Accessed June 15, 2018.

17. Handly, B. (2017, December 20). A New Toy Story: How Retailers Can Use 'Geo-Conquesting' to Stay Relevant. *Forbes.* Available at https://www.forbes.com/sites/forbestechcouncil/2017/12/20/a-new-toy-story-how-retailers-can-use-geo-conquesting-to-stay-relevant/#40452aad552e. Accessed June 15, 2018.

18. Aziza, B. (2018, January 8). Big Data: Amazon, Google, Microsoft, the Cloud and Other 2018 Trends. *Forbes.* Available at https://www.forbes.com/sites/cio-central/2018/01/08/big-data-amazon-google-microsoft-the-cloud-and-other-2018-trends/#193cc9ef2ba1. Accessed July 9, 2018.

19. Hirschler, B. (2018, March 1). Big Pharma, Big Data: Why Drugmakers Want Your Health Records. *Reuters.* Available at https://www.reuters.com/article/us-pharmaceuticals-data/big-pharma-big-data-why-drugmakers-want-your-health-records-idUSKCN1GD4MM. Accessed July 9, 2018.

20. Hughes, D., & Poletti-Hughes, J. (2016, October 21). How Pharmaceutical Companies Profits from Drugs for Rare Diseases. *The Conversation.* Available at https://theconversation.com/how-pharmaceutical-companies-profit-from-drugs-for-rare-diseases-65266. Accessed July 9, 2018.

21. Stone, K. (2017, December 11). The Most Expensive Prescription Drugs in the World. *The Balance.* Available at https://www.thebalance.com/the-8-most-expensive-prescription-drugs-in-the-world-2663232. Accessed July 9, 2018.

22. Verhoef, P. C., Kooge, E., & Walk, N. (2015). *Creating Value with Big Data Analytics: Making Smarter Marketing Decisions.* London: Routledge.

23. Redman, T. C. (2017, June 15). Does Your Company Know What to Do with All Its Data? *Harvard Business Review.* Available at https://hbr.org/2017/06/does-your-company-know-what-to-do-with-all-its-data. Accessed July 9, 2018.

24. Morey, T., Forbath, T., & Schoop, A. (2015, May). Customer Data: Designing for Transparency and Trust. *Harvard Business Review, 93,* 96–105.

25. Bloomberg. (2018, February 20). The Car of the Future Will Sell Your Data. *Fortune.* Available at http://fortune.com/2018/02/20/car-future-sell-data-telenav/. Accessed August 20, 2018.

26. McKinsey Analytics. (2017, December). *Fueling Growth Through Data Monetization.* Available at https://www.mckinsey.com/business-functions/mckinsey-analytics/our-insights/fueling-growth-through-data-monetization. Accessed July 9, 2018.

27. Davenport, T. H., & Patil, D. J. (2012, October). Data Scientist: The Sexiest Job of the 21st Century. *Harvard Business Review, 90,* 70–76.

28. McKinsey Global Institute. (2016). *The Age of Analytics: Competing in a Data-Driven World.* Available at www.mckinsey.com/mgi. Accessed June 12, 2018.

29. *The Edge.* (2012, August 30). Reinventing Society in the Wake of Big Data: A Conversation with Alex 'Sandy' Pentland. Available at https://www.edge.org/conversation/reinventing-society-in-the-wake-of-big-data. Accessed July 9, 2018.
30. Martin, K. D., Borah, A., & Palmatier, R. W. (2017, January). Data Privacy: Effects on Customer and Firm Performance. *Journal of Marketing, 81,* 36–85.
31. Scheiber, N. (2017, September 28). Facebook's Ad-Targeting Problem, Captured in a Literal Shade of Gray. *New York Times.* Available at https://www.nytimes.com/2017/09/28/technology/facebook-ads.html. Accessed July 9, 2018.
32. Ganatra, R. (2018, January 18). 22 Retail Industry Predictions for Brick-and-Mortar Stores in 2018. *Forbes.* Available at https://www.forbes.com/sites/rganatra/2018/01/18/22-retail-industry-predictions-for-brick-and-mortar-stores-in-2018/#1f05739f7000. Accessed July 9, 2018.
33. For exceptions see Haenlein, M., & Kaplan, A. M. (2009, January–February). Unprofitable Customers and Their Management. *Business Horizons, 52,* 89–97; and Shin, J., Sudhir, K., & Yoon, D.-H. (2012). When to 'Fire' Customers: Customer Cost-Based Pricing. *Management Science, 58*(5), 932–947.
34. Bone, S. A., Christensen, G. L., & Williams, J. D. (2014, August). Rejected, Shackled, and Alone: The Impact of Systemic Restricted Choice on Minority Consumers' Construction of Self. *Journal of Consumer Research, 41,* 451–474.
35. Safdar, K. (2018, April 4). The Stores That Track Your Returns. *Wall Street Journal.* Available at https://www.wsj.com/articles/the-stores-that-track-your-returns-1522843201. Accessed July 9, 2018.
36. Bowles, N. (2018, June 23). Thermostats, Locks and Lights: Digital Tools of Domestic Abuse. *New York Times.* Available at https://www.nytimes.com/2018/06/23/technology/smart-home-devices-domestic-abuse.html. Accessed July 9, 2018.
37. Gualtieri, M. (2016). *Artificial Intelligence: What's Possible for Enterprises in 2017.* Cambridge, MA: Forrester Research.
38. Power, B. (2017, May 31). How Harley-Davidson Used Artificial Intelligence to Increase New York Sales Leads by 2,930%. *Harvard Business Review.* Available at https://hbr.org/2017/05/how-harley-davidson-used-predictive-analytics-to-increase-new-york-sales-leads-by-2930. Accessed July 9, 2018.

5

Inoculating Against Customer Vulnerability

Introduction

This chapter offers defensive strategies to protect customers from vulnerability and thereby guard the firm against financial harm and customer defection. We compile best practices from academic studies, business advice, expert insights, and our own extensive research with companies and customers. Rather than dealing with the technical details of cybersecurity, which is the purview of IT experts, we instead study and review inoculation strategies that can protect the firm by establishing sound data privacy "marketing" practices that are designed essentially to benefit customers. Companies can create powerful shields to guard against the harms of privacy failures simply by keeping customers' best interests as the core of everything they do. We acknowledge that there are many sophisticated approaches for keeping data safe, but our goal is to consider the ways data are collected and managed by the company's marketing team, because today's marketers must take responsibility to consider how they gather and use data in the context of establishing and offering customer value. We also continue to emphasize the value of transparency and control, as well as creating a company culture that explicitly prioritizes minimizing customer harm. Such companies can realize important benefits to their own bottom lines and create an effective defense against the myriad harms of data breaches and other possible privacy failures.

© The Author(s) 2019
R. W. Palmatier and K. D. Martin, *The Intelligent Marketer's Guide to Data Privacy*,
https://doi.org/10.1007/978-3-030-03724-6_5

We begin with an overview of how privacy failures affect companies, both directly and indirectly. Most marketers are aware of the dangers posed to their own companies; the harms that can occur through spillovers when a close rival firm experiences a breach are less intuitive but equally threatening. In this discussion, we make the case for why no company can be complacent when it comes to adopting and implementing strong customer data privacy practices. Three strategies should work together to defend the company from privacy failure harms: (1) practicing data minimization, (2) eschewing a data-driven culture in favor of a customer-learning culture, and (3) empowering customers with transparency and control. Together, these strategies go beyond defending the firm from privacy lapses, because they reinforce customer relationships for long-term advantage.

> To defend their company from privacy failure harms, marketers should apply three strategies in combination: (1) practicing data minimization, (2) eschewing a data-driven culture in favor of a customer-learning culture, and (3) empowering customers with transparency and control.

No Company Is Safe

Recall that in Part I, we presented evidence that customers feel vulnerable, simply because companies have access to their personal information. For this reason alone, marketers should take a careful look at how they approach customer data and their uses. Privacy failures of all types occur at companies with almost predictable regularity—ranging from small, internal breaches with minor risks to vast cyberattacks that can compromise millions of records. Regardless of the scale or scope though, greater IT security and technological sophistication simply are proving insufficient barriers to cybersecurity threats. Even after a breach, newly reinforced cybersecurity measures do little to deter future attacks.

Rather, after a company experiences a substantive data breach, the likelihood that it will suffer a subsequent breach is greater than 27% [1]. This statistic is especially alarming when we realize that more than 1500 U.S. organizations were affected by some form of data breach in 2017 alone [2]. Data breaches that affect organizations of all types now happen with such high probability that experts refer to them as the "third certainty ... after death and taxes" [3]. The average firm costs associated with these events approach $4 million, or $141 per record, in addition to longer term stock price and other financial performance damages. Furthermore, the losses are not limited to the breached company; in studying hundreds of them, we have realized that data breaches have powerful ripple effects that can either harm (due to *spillover effects*) or help (due to *competitive effects*) a company's competitors.

Spillover Effects

A key detrimental impact of a data breach on a company results from the anticipation of negative customer responses and a perception of insufficient data protection by the firm. These forces combine to damage firm performance, as reflected in its stock returns. That is, the firm's stock price likely decreases when an efficient market anticipates lost future sales to existing customers, increased difficulty acquiring new customers, or potential legal and recovery costs. When these negative events garner unfavorable publicity, the influence also can spread to rival firms, through a "guilt-by-association" effect, such that crises can harm the firms that represent close rivals to the affected firm [4]. These negative spillover effects occur because customers believe the nature or root cause of the crisis is endemic to the entire category or industry [5]. After a data breach by a focal firm, customers of rival firms may feel more vulnerable, which creates a cascade of actions and negative spillover onto the rival firm's performance, due to anticipation in the stock market. Smaller breaches in particular may signal a higher potential vulnerability to similar hacking by others in the industry, including close competitors. A breach of this nature can signal industry-wide problems or weaknesses that make others look susceptible too.

Competitive Effects

Yet brand scandal research also offers an alternative perspective: A data breach event may create a positive competitive effect for the closest rival that mitigates or offsets the negative spillover damage. If the breach creates severe negative publicity, customer backlash, and financial harm to the focal firm, its customers might switch to a rival. Customers often shift away from a firm experiencing a brand crisis, and this switch ultimately may be permanent [6]. A rival firm then gains sales and profits from new customers, which improves its financial performance.

Our research shows that severity, or the number of customers affected by a breach is the key to understanding whether close rivals will be harmed or helped by their competitors' bad fortune. As the number of customers harmed increases, stock market effects for the firm's rivals go from negative (due to spillover effects) to positive, because the competitive effects come to dominate. As we noted, smaller breaches may create an impression of being common throughout the industry, but large data breaches create impressions of extreme trouble primarily for the focal firm. During large data breaches, customers' desire to leave the breached firm also become more salient. The resulting switching behavior thus benefits the breached firm's competitors, as reflected in their stock returns.

The good news for marketing managers and companies is that they are not powerless in the face of data breaches. They have access to actionable strategies to protect or "inoculate" them from breaches, whether at their own firm or a close rival firm. Then the company also inoculates the customer from the harms that result from a lack of company transparency and capacity to control the use of their personal information. As we detail in the remainder of this chapter, marketers can defend against privacy failures by better managing their orientation and approach to collecting customer data.

> The good news is that marketing managers and companies are not powerless when it comes to data breaches affecting their competitors. They have actionable strategies to protect or "inoculate" them from breaches at their own firm or at a close rival firm.

In Praise of Data Minimization

The business software giant SAP estimates that up to 73% of all data collected by companies are never used [7]. If this statistic is accurate, the concept of data minimization may represent a welcome way to rethink the firm's approach to data management. When they embrace *data minimization*, firms willingly limit the type and amount of personal information they collect about their customers. Although data minimization is an increasing topic of discussion, largely due to its inclusion in the GDPR principles, prior research for years has advocated its practice as a way to protect privacy [8]. In our own research, we find support for the foundational concepts of data minimization. In studies with thousands of customers, we recognize that people experience strong feelings of vulnerability simply from firms accessing their data. Data minimization directly addresses in this challenge and alleviates customers' felt vulnerability from mere data access. Moreover, data minimization offers additional firm benefits.

Formal definitions of data minimization cite (1) limited firm opportunities for collecting customer personal information; (2) within those opportunities, constraining the categories of customer information collected; (3) condensing the depth and detail of customer information collected within those categories; and (4) reducing the amount of time the firm keeps customer personal information in its possession. These practices are relevant, and they largely underpin the core ideas in GDPR's explication of data minimization (see Chapter 3 for more information on GDPR). The GDPR guidelines—specifically, Article 5, which details principles relating to processing of personal data—mandate that customer personal information and data should be collected sparingly. The specific policy item devoted to data minimization reads, personal data will be "adequate, relevant and limited to what is necessary in relation to the purposes for which they are processed" [9]. This premise dovetails with initiatives to limit data collection to the purposes for which it is specified (rather than gathering additional, unrelated information), as well as mandates a limit to storage time so that firms do not retain personal data for longer than they need them. Together, these principles tell firms that they only may gather the minimum amount of relevant information needed to answer a specific question. Even then, the data may not be stored in perpetuity but rather must have a finite lifespan at the firm.

These constraints may appear to be an impediment to successful customer research and analytics, but we argue that the principles of data minimization also can be an important guideline for marketers. Specifically, principles of data minimization that limit unnecessary data collection and promote greater relevance force the firm to focus its efforts. It puts the responsibility on the market research or customer analytics team to specify, what question do we need to answer? Data minimization accordingly allows market researchers to return to best practices of the field, advocated in textbooks for decades [10]. A sequential research process, as advocated in marketing pedagogy, always begins with formulating a specific question. When marketers are forced to be precise in their question of interest, it helps them obtain more accurate and appropriate information. The narrow scope also allows the firm to be more intentional in its data collection efforts. The ease, accessibility, and abundance of big data have pushed many marketers away from stringent best practices, in which methodical research question formulation leads to subsequent testing.

This sharper focus not only makes research efforts more purposeful and strategic, but it also prompts more efficient data collection, which can be far more cost-effective than broad-based approaches designed to gather as much data as possible. Data minimization limits waste and prevents distraction by the research and analytics teams. Therefore, data minimization principles encourage a better use of firm resources and greater marketing accountability. Marketers can pinpoint customer needs and wants, then create valued products, services, and experiences. Even companies that are proverbially swimming in data extoll the virtues of data minimization. Seth Stephens-Davidowitz, former Google employee and author of the book *Everybody Lies*, which explores the usefulness of big data, reinforces these ideas: "At Google, major decisions are based on only a tiny sampling of their data. You don't always need a ton of data to find the right insights. You need the right data" [11]. Data minimization thus represents a strong, strategic approach to formulating focused questions, effectively and efficiently finding the right information, and using the information to solve deep and important marketing problems. In the meantime, marketers have guarded their customers against potential harms and unforeseen misuses of their data—an important defensive approach that further inoculates the firm from harm.

From a Data-Driven Culture to a Customer-Learning Culture

A key impediment to data minimization though, and to inoculating against customer vulnerability more generally, is the increasing call for companies to embrace a data-driven culture. A recent survey with top executives reports that 85% of firms express a strong desire to create a data-driven culture, and 37% report having already achieved this organizational type [12]. Simultaneously, the role of Chief Data Officer (CDO), primarily tasked with security issues, compliance, and regulatory responses, is growing and becoming more proactive. In this sense, CDOs may be at the forefront of data monetization and innovative efforts to expand and exploit this increasingly important source of firm revenue.

In no way do we mean to advocate against the myriad benefits that can be realized by integrating data analytics and insights within a company. As we noted in Chapter 4, the ease of access, low cost, and flexible nature of customer data promotes an experimental mindset and allows marketers to explore innovative ways to delight customers. These are critical advantages that should be part of every modern organization. However, we advocate an approach that requires companies consider the big picture—including the real people who are their customers—at the forefront of all decision making.

The following quote from Zoher Karu, Vice President of Global Customer Optimization and Data at eBay as interviewed by McKinsey Analytics, supports this corporate mindset. "One of the biggest challenges is around data privacy and what is shared versus what is not shared. And my perspective on that is consumers are willing to share if there's value returned. One-way sharing is not going to fly anymore. So how do we protect and how do we harness that information and become a partner with our consumers rather than kind of just a vendor for them?" [13]. In further support, a recent *AdWeek* article, produced in partnership with Accenture, advises that despite their vast access to troves of data, companies should embrace and emphasize a corporate responsibility culture, even if a data-driven mindset informs their marketing decisions [14].

Perhaps a better way to frame current calls for data-driven culture is to focus on what data do for the company and how they help the company better serve customers. We classify these ideas into several categories in Exhibit 5.1. Some executives already have made this transition in the organizational mindset, from data-driven to creating a learning culture. The retired American Express CEO Ash Gupta noted of his firm's push to forge a data-driven culture, "Good data-driven organizations have a 'test-and-learn' culture and encourage experimentation that can challenge long-held beliefs. These organizations promote listening and learning, and adequately synthesize it with the institutional knowledge and experience" [15]. American Express focuses less on the data themselves, and more on what it can learn from data-enabled technologies. The company can boast new initiatives as a result, including a successful small business lending program. What made such new programs viable was a learning culture mindset, coupled with a strong, untapped customer need, which American Express identified through data and analytics. It committed to serving customers above all though, never deviating from this promise, or as Gupta describes, "keeping the interests of the customer first in everything we do and being a good citizen in the communities which we serve." Not surprisingly, American Express also ranks toward the top of *Fortune* 500 firms when it comes to protecting customer privacy [16].

Other companies proclaim the benefits of a customer-learning culture too. For example, Victor Nilson, Senior Vice President of Big Data at AT&T, explains, "We always start with the customer experience. That's what matters most. In our customer care centers now, we have a large number of very complex products. We take the complexity out and turn it into something simple and actionable" [17]. At the warehouse chain Costco, customer

Exhibit 5.1 Comparing the data-driven culture and the customer-learning culture

	Data-driven organization	Customer-learning organization
Overarching data philosophy	Data drive what the firm does; data are the primary source of insights	Data support the company's ongoing need to learn about customers, competitors, and the environment
Strategic decision making	Driven by what the data reveal	Driven by customer-oriented goals for long-term advantage; supported by data-analytical insights
Big picture view of the customer	Source of data (and profit)	Partners in value creation
Experimental mindset	Embraced for financially optimal outcomes; often short-term	Embraced for customer optimal outcomes; often long-term
Approach to customer information	Data maximization	Data minimization
Data tactics: access	Permission only when needed	Explain to customers why providing their information can help them
Data tactics: use	Data purpose is determined after access; data analyzed in any way that helps the firm	Data purpose is determined prior to access; data analyzed in ways for which they were intended
Data tactics: protection	Data are stored indefinitely; company uses whatever protections necessary	Data are housed only as long as needed for specific question; need for protection thus is minimized
Data tactics: monetization	Customer data are focal profit sources, sold to third parties as feasible	Customer data are collected for focused questions; their dedicated nature make them unviable profit source

data inform its product assortment decisions, but they also led the retailer to offer post-sale customer services unlike any other warehouse retailer. Costco's customer data management system helped the company identify people who had purchased some specific produce that was at risk of a potentially dangerous *listeria* outbreak. It quickly notified the customers by phone about the possible contamination, then followed up with email communication checks. The company's systems are so accurate that Costco has partnered with the U.S. Centers for Disease Control and Prevention to pinpoint sources of unrelated food contamination events [18]. Renowned for protecting customer privacy though, Costco also makes sure that it empowers people with transparency about how their personal information will be used and giving them a say in those uses. Empowering customers with data transparency and control has relevant implications for firms; not having those practices in place also has ramifications, as we discuss in the next section.

Empowering Customers with Transparency and Control

Many marketing experts have a sense for engaging with customers in a way that those customers appreciate, so they respond favorably by willingly providing their personal information. As Phil Sutcliffe, Head of Offer and Innovation at Kantar TNS U.K., notes, "People talk about trust in brands in a similar way to which they talk about trust in people—it's quite emotive—so it's important for brands to embody those human qualities and to be quite real and vulnerable and transparent in terms of how they are communicating" [19].

Our previous research [20] demonstrates that the two firm privacy practices that customers crave most are (1) data transparency and (2) customer control of that data. Across dozens of studies, transparency and control emerge as focal determinants of customers' willingness to trust the firm, feel less violated, and ultimately continue to purchase from the firm without defecting to competitors. We tested these attributes among many others, including value, reciprocity, fairness, ethics, and other cognitions and emotions often linked to company data use. Again and again, transparency and control rose to the top, surpassing any other actions firms might take to ensure customers privacy through various behaviors and practices. What is more, for companies of all types, these two key facets of customer empowerment are readily implementable and, as we describe, can save them millions if a privacy lapse eventually does occur.

Breaking Down Transparency and Control

Transparency and control already exist (or fail to exist) in company privacy practices in many ways. Transparency refers to the company's willingness and ability to clearly explain to customers how it is collecting, using, sharing, or protecting their data. Transparent privacy practices tell customers what specific information the firm captures and how it uses that information (e.g., IP address, search history, promotions, sold to third parties). Transparency thus implies that customers have knowledge of the nature and scope of data the firm possesses and how those data will used. By offering transparency, firms can avoid the "creepiness" that customers feel when a firm uses their data to create highly targeted advertising. Customers who are aware of what data are being collected and what they will be used to do become less suspicious of the firm and express higher levels of trust.

Control instead refers to allowing customers to decide how their data will be used and shared. It is achieved by giving customers opportunities to opt out of firm data practices (e.g., promotion, sharing with partners, cross-selling). When companies offer customers control in their data privacy practices, they make it easy for people to access their personal information and to determine how it can be used. If customers identify errors in their information, they can be readily corrected. If accounts at one service provider are linked with others, the customer can decide how to manage those connections. Customers can decide how, when, or if they would like the company to communicate with them. Bestowing control also means that when the customer

> Our research with customers, together with examinations of firm stock market performance following a data breach, show that transparency and control effectively empower customers. Empowered customers do not punish firms for data breaches. Instead, they are more willing to share information and more forgiving of data privacy breaches, so they remain loyal after the fact.

sets her or his parameters for control, the company honors and upholds those wishes.

Customer Empowerment Outcomes

Consumer psychology research defines customer empowerment as a combination of knowledge and control. Formally, *empowerment* involves "strategies or mechanisms that equip people with sufficient knowledge and autonomy to allow them to exert control over a certain decision" [21]. Such research evinces the logical efficacy of transparency and control with regard to inoculating customers from vulnerability in data privacy settings. Our research with customers, as well as examinations of firm stock market performance following a data breach, show that together, transparency and control effectively empower customers. Empowered customers do not punish firms for

Empowered Customer High Transparency, High Control	Blind Autonomy Low Transparency, High Control
Empowered customer has a clear picture of how data is used and has some say in controlling it. Evidence shows these firms endure minimal monetary losses during a data breach.	Customer has the autonomy to control data preferences, but little idea of how data actually is used. Increasing transparency with high control can save a firm about $50 million in a data breach.
Percentage of Fortune 500 that does it: 10%	Percentage of Fortune 500 that does it: 10%
Privacy Grade: A	Privacy Grade: B-
Look but don't Touch High Transparency, Low Control	**Privacy Deficient** Low Transparency, Low Control
Data use practices are clearly communicated, but customer but has no ability to manage it. Increasing control when data practices are transparent can create firm savings during a breach of nearly $60 million.	Perceived as sneaky approach to privacy. Customer has no knowledge how data is used, but can't control it anyway. These firms are at greatest risk during a breach. Expected firm losses from a breach near $40 million.
Percentage of Fortune 500 that does it: 40%	Percentage of Fortune 500 that does it: 40%
Privacy Grade: D	Privacy Grade: F

Exhibit 5.2 Privacy scorecard, who is best protected?

data breaches; they are more willing to share information and even are more forgiving of data privacy breaches, such that they remain loyal in the aftermath. Customers of firms that offer high transparency and control feel less violated by big data practices, are more trusting, provide more accurate data, and generate positive word of mouth (WOM). Firms that score high on these two dimensions also enjoy buffers from stock price damage due to data beaches, whether by their own firm or by their closest rivals. Yet by our accounts, only about 10% of *Fortune* 500 firms fit this profile, as assessed by the Privacy Scorecard summarized in Exhibit 5.2.

We thus sought to understand just how much protection *Fortune* 500 firms could achieve, by studying their use of transparency and control in their privacy policies. Our research team combed the privacy policies of all *Fortune* 500 firms to gain insights [22]. Privacy policies provide important communication tools, because firms agree legally to abide by them. Regardless of what a firm might say about data privacy in other channels, it must put into practice the promises it formally documents in its privacy policy. When customers are in doubt about their personal information, firm responses commonly refer them back to the privacy policy. According to a recent review of data privacy research in marketing [23], customers have a good idea of firms' data practices, as summarized in firms' privacy policies—even if they do not read the entire, actual privacy policy. Because firms' privacy-related behaviors are so salient to customers, it is critical that firms think carefully about them.

Some firms provide high levels of data transparency and control and thus would be protected, should a data breach or other privacy failure occur. Likewise, these companies tend to be shielded from the spillover effects that emerge when a close competitor experiences a data breach. That is, companies can avoid harm to themselves, as well as possible spillover effects from a close rival, by clearly conveying what information they capture and how, while also offering customers substantial control over how they share and use that information. Considering this clear evidence, it seems surprising that so many firms fail to provide their customers with either dimension. Ultimately, these firms are at the greatest risk of financial harm and customer defection in response to a privacy failure. When evaluating the financial performance of companies after a data breach, we find that firms that exhibit low transparency and control lost millions more than firms just like them that empowered their customers.

Across the scorecard, we find many instances of firms that appear to offer either transparency or control, but not both, to customers. Consistently in our studies, we note that when provided with high transparency but low control, customers perceive more violations and lower trust. This approach is poorly received, because customers know about data uses and sharing but have no means to control those practices. It is dangerous for firms to tell customers how they will be collecting data, without also providing them with some say over those uses. If they lack control, customers are left to worry about the various potential applications of their data—applications made salient by their transparency. Knowledge by itself has mixed effects as a vulnerability suppressor. Hence, if firms intend to reveal their data use practices to customers, they also need to provide customers with some element of control over the information.

Just providing control without requisite knowledge also fails to empower customers sufficiently. For example, some firms provide autonomy through control but fail to explain how that autonomy actually maps to their data privacy practices (low transparency). The combination of low data use transparency and high control creates a situation of uninformed autonomy. Customers have the ability to update their preferences, so they respond favorably, even if their opt-in and -out choices are somewhat blind, without full knowledge of how the firm uses their information [24]. Collectively, these contrasts suggest a powerful managerial tool for generating positive firm outcomes. The amount of customer control provided might not need to reach full and total autonomy; rather, sufficient levels of control that are meaningful to customers may produce the desired mitigating effects. By allowing customers to opt in or out of various data practices, firms can promote greater overall customer willingness to provide personal information.

Finally, firms that neither tell customers about how they use their data nor offer any control are at the greatest risk for harm. For these firms, a localized data breach or that of a close competitor poses serious threats to their financial performance. Yet our privacy coding identified an overwhelming 80% of *Fortune* 500 firms that fall into one of these two highest risk categories. In our study of data breach events and their immediate impact on stock prices, firms that fail to explain data privacy practices have a 1.5 times greater drop in stock price than firms with high transparency; firms that provide customers high control over their data experience no significant change in their stock price. Enabling informed, autonomous customers thus inoculates firms from damage during severe events such as data breaches, whether of their own or their rivals.

Keeping the Customer Relationship Focal: Advice to Managers

Customer data practices help marketers identify and understand customers and segments, but these same practices can create vulnerability throughout the customer cohort. Our findings suggest that firms can inoculate themselves from customer vulnerability–induced performance harms by adopting a more tempered approach to data and analytics initiatives. Marketers should consider their approaches to data management carefully, as well as embrace data minimization and a customer-learning culture to avoid negative effects. We highlight the power of customer empowerment, using transparency and control as ways to suppress customer vulnerability and enhance the positive aspects of the customer–company relationship. Empowering customers is critical for companies of all types. This point is especially pertinent, considering our spillover effect findings. Even non-compromised firms can suffer substantial financial performance detriments when a close competitor has a breach.

Empowering customers by providing transparency and control is the hallmark of a healthy relationship. Customer empowerment is a natural outgrowth of keeping the customer relationship focal to everything the firm does, as we review in our discussion of organizational culture. More pragmatically, transparency and control combine to prevent the vulnerability that customers feel from the company's data uses, so it also limits the harm to firm performance. Across our various studies, we find that the most potent vulnerability-suppressing combination provides customers with clear transparency and control over their personal information. Strong transparency and control reduce the spread of negative WOM, deter switching, and suppress negative stock price effects. Customer empowerment represents a strong promoter of company trust. Noting recent evidence that nearly 70% of customers do not trust companies with their personal information [25], empowerment offers an effective way to support customer relationships and overcome negative implications.

Ultimately, firms can protect against damage to their own performance by strengthening their own data privacy practices, which also can shield them against potential spillovers from competitors' privacy failures. However, such changes need to be meaningful and penetrate beyond surface-level company practices. Marketing, IT, and legal teams need to collaborate to develop customer-friendly, legally implementable, and technologically actionable strategies that leave customers empowered. Aspects of empowerment that matter most include explaining clearly the ways in which the firm will access, use, share, and protect customer information. Being informed must go hand-in-hand with customer autonomy, to allow them to adjust the rules for using their data. A failure to do so leaves a firm susceptible to risk from multiple harms.

References

1. Ponemon Institute. (2017). *Cost of Data Breach Study: Global Overview.* Ponemon Institute and IBM Security Research Report.
2. Identity Theft Resource Center. (2018). *Annual Data Breach, Year-End Review.* Available at www.idtheftcenter.org. Accessed July 14, 2018.
3. See, for example, Zaretsky, R. (2015, June 2). There Are Three Certainties in Life: Death, Taxes, and Data Breaches. *Tax Policy Center of the Urban Institute and the Brookings Institute.* Available at https://www.taxpolicycenter.org/taxvox/there-are-three-certainties-life-death-taxes-and-data-breaches. Accessed July 14, 2018.
4. Borah, A., & Tellis, G. (2016, April). Halo (Spillover) Effects in Social Media: Do Product Recalls of One Brand Hurt or Help Rival Brands? *Journal of Marketing Research, 53*, 143–160.
5. Cleeren, K., van Heerde, H. J., & Dekimpe, M. G. (2013, March). Rising from the Ashes: How Brands and Categories Can Overcome Product-Harm Crises. *Journal of Marketing, 77*, 58–77.
6. Roehm, M. L., & Tybout, A. M. (2006, August). When Will a Brand Scandal Spill Over, and How Should Competitors Respond? *Journal of Marketing Research, 43*, 366–373.

7. Millar, B. (2018, April 19). Big Data Is a Sham. *Fast Company*. Available at https://www.fastcompany.com/90168426/big-data-is-a-sham. Accessed July 11, 2018.

8. Pfitzmann, A., & Hansen, M. (2010). *A Terminology for Talking About Privacy by Data Minimization: Anonymity, Unlinkability, Undetectability, Unobservability, Pseudonymity, and Identity Management*. University of Dresden Working Paper.

9. GDPR. Article 5. *Principles Relating to Processing of Personal Data*. Available at https://gdpr-info.eu/art-5-gdpr/. Accessed July 11, 2018.

10. See, for example, Burn, A. C., Veeck, A., & Bush, R. F. (2017). *Marketing Research* (8th ed.). Boston: Pearson.

11. Stephens-Davidowitz, S. (2017). *Everybody Lies: Big Data, New Data, and What the Internet Reveals About Who We Really Are*. New York: HarperCollins.

12. NewVantage Partners. (2017). *Big Data Executive Survey 2017*. Boston: NewVantage Partners LLC.

13. McKinsey Analytics. (2016, April). How Companies Are Using Big Data and Analytics. *McKinsey & Company*. Available at https://www.mckinsey.com/business-functions/mckinsey-analytics/our-insights/how-companies-are-using-big-data-and-analytics. Accessed August 27, 2018.

14. St. Louis, M. (2018, April 3). 3 Ways Businesses Can Use Big Data Responsibly. *AdWeek*. Available at https://www.adweek.com/digital/3-ways-businesses-can-use-big-data-responsibly/. Accessed July 12, 2018.

15. Bean, R. (2018, March 15). How American Express Excels as a Data-Driven Culture. *Forbes*. Available at https://www.forbes.com/sites/ciocentral/2018/03/15/how-american-express-excels-as-a-data-driven-culture/#116fa6ed1635. Accessed July 12, 2018.

16. Martin, K. D., Borah, A., & Palmatier, R. W. (2018, February 15). A Strong Privacy Policy Can Save Your Company Millions. *Harvard Business Review*. www.hbr.org.

17. McKinsey Analytics. (2016). How Companies are Using Big Data and Analytics. April Interviews. Available at https://www.mckinsey.com/business-functions/mckinsey-analytics/our-insights/how-companies-are-using-big-data-and-analytics. Accessed July 12, 2018.

18. Wilson, J. (2014, July 24). More Stores Affected by Listeria Fruit Recall. *CNN*. Available at https://www.cnn.com/2014/07/22/health/costco-fruit-recall/index.html. Accessed July 12, 2018.

19. As quoted in O'Reilly, L. (2018, June 19). The Future of Digital Marketing in a Data-Privacy World. *The Wall Street Journal*, R1–R2.

20. Martin, K. D., Borah, A., & Palmatier, R. W. (2017, January). Data Privacy: Effects on Customer and Firm Performance. *Journal of Marketing, 81*, 36–85.

21. Quote from Camacho, N., de Jong, M., & Stremersch, S. (2014). The Effect of Customer Empowerment on Adherence to Expert Advice. *International Journal of Research in Marketing, 31*(3), 293–308. See also the work of Albert Bandura, e.g., Bandura, A. (1989). Human Agency in Social Cognitive Theory. *American Psychologist, 44*, 1175–1184.

22. Martin, K. D., Borah, A., & Palmatier, R. W. (2018, February 15). A Strong Privacy Policy Can Save Your Company Millions. *Harvard Business Review*. www.hbr.org.
23. Martin, K. D., & Murphy, P. E. (2017, March). The Role of Data Privacy in Marketing. *Journal of the Academy of Marketing Science, 45*, 135–155.
24. Martin, K. D., Borah, A., & Palmatier, R. W. (2017). How Should Marketers Manage Data Privacy? *AMA Scholarly Insights*. Available at https://www.ama.org/resources/Pages/how-should-marketers-manage-data-privacy.aspx.
25. Rose, J. (2018). *The State of Consumer Privacy and Trust in 2017: Fear and Hope*. GIGYA Research Report. www.gigya.com.

6

Privacy Failures and Recovery Strategies

Introduction

Data breach events are reported on a regular basis. The nonprofit Privacy Rights Clearinghouse calculates there have been nearly 1 trillion compromised personal records from data breaches to date, since it began tracking publicly announced events in 2005. What makes this number particularly staggering is that the scope of most data breach events is unknown, and the extent and number of compromised records simply cannot be quantified [1]. The Identity Theft Resource Center, publishing in conjunction with the U.S. government, also tracks thousands of data breaches and estimates that they have cost firms hundreds of millions of dollars in lost sales and recovery costs, even before accounting for their damages to long-term performance and reputations [2].

Data security experts define a data breach as "a confirmed incident in which sensitive, confidential or otherwise protected data has been accessed and/or disclosed in an unauthorized fashion" [3]. Data breaches typically involve a customer's financial information, credit card details, personal health or medical information, or any other personally identifiable information. Although all 50 U.S. states now mandate that firms notify customers when a data security breach has occurred, there is considerable flexibility in the way such acknowledgment can happen [4]. For example, firms are subject to widely varying parameters of notification, including when the breach must be disclosed, how it may be communicated, and what particular information should be conveyed in the message [5]. To date, there is no clear

© The Author(s) 2019
R. W. Palmatier and K. D. Martin, *The Intelligent Marketer's Guide to Data Privacy*,
https://doi.org/10.1007/978-3-030-03724-6_6

policy guidance regarding the optimal strategies for recovering from firm data breaches.

Academic research indicates a negative effect of data breaches on firm performance. As we have detailed in previous chapters, data breaches have clear, negative effects on the stock price of the affected focal firm but also can spill over to close competitors [6]. But how should they recover from these events? Business practitioner advice typically takes the perspective that the goal should be enhancing information technology (IT) security outcomes and solutions, with an emphasis on strengthening the firm's IT security to guard against future intrusions [7]. Some investigations also address the extent to which firms may experience legal sanctions or litigation following a data breach (see Chapter 3 for more information on legal ramifications and liability) [8]. According to a study of 230 federal data breach lawsuits, when consumers suffer financial harm, breached firms are 3.5 times more likely to face litigation. Financial harm to customers also increases the likelihood that the breached firm will settle the lawsuit, by approximately 30%. Yet in all the research exploring data breach occurrences, we find little advice about how to preserve customer relationships, convey information about the breach, or offer compensation in a way that prevents defection. Marketing managers who primarily seek to reconnect with customers following a data breach lack clear guidance for repairing their relationships.

We attempt to fill this void. We explicitly designed this chapter to explore recovery from data breaches in a way that speaks directly to marketers versus lawyers or IT specialists. Related marketing crises help set the stage for managing a data breach in ways that minimize customer harm. Examples from real company data breaches and subsequent recovery efforts help illuminate key issues. However, careful examination of data breach events, relative to other negative customer–firm interactions, suggests that learning about a data breach represents an extreme and uncertain form of customer bad news, so it requires special, dedicated crises communication considerations. We draw from research and best practice prescriptions for conveying truly bad news in medical settings to better understand the various drivers and mechanisms at play when firms inform customers that their personal information—the ingredients criminals can use to fraudulently adopt their identity—has been compromised. We integrate these perspectives with crisis communication theory [9] and thus develop specific approaches to informing and assisting customers following a breach. As such, this chapter is designed to help marketers identify and implement important delivery and content elements of data breach recovery strategies to reduce negative effects, in a way that specifically addresses the unique data breach context.

Data Breach as a Marketing Crisis

Various suggestions exist for regaining control of IT systems and minimizing the scope of compromised data using a firm's existing information security infrastructure. But advice about how to recover the customer base following such an event is far less extensive. Neither industry best practices nor uniform regulatory guidelines are available when it comes to the manner in which a firm should respond to customers following a data breach, which often puts marketers in unmapped terrain [10]. The absence of practical advice, coupled with a dearth of academic research, produces very little understanding overall of the effective means by which companies can restore customer relationships following a data breach. Marketers in turn wind up formulating customer data breach responses blindly. Such uninformed reactions are problematic in themselves, because a particular response strategy can be expensive (e.g., providing credit protection services to customers) or unintentionally harmful, such as when they create needless customer worry (e.g., telling everyone about a minor data breach that affected very few customer records).

Business and popular press are replete with examples of how firms have failed to recover. Target was criticized for deploying its recovery strategy too late; various sources leaked the relevant details of its massive data breach before the company ever even acknowledged something had occurred. Once it did acknowledge the breach publicly, Target's CEO effectively provided a personal apology and an offer of a year of free credit monitoring to customers [11]. However, many analysts found Target's added offer of a 10% discount on a future purchases an unsavory attempt by the retailer to benefit from the harm [12]. When eBay experienced a massive data breach, criticisms instead focused on its failure to communicate directly with customers. Instead, the retailer posted a small notice on a little-used alternative corporate website with basic advice to customers to change their passwords [13]. The craft retailer Michael's drew criticism for the long announcement delay following its large breach, but also for casting itself as the victim, describing the event as an "unprecedented attack using sophisticated malware" rather than acknowledging the damage to customers [14]. Even this small sample of examples of what firms should not do following a breach establishes the critical need for greater insights.

We argue that the best practices for data breach recovery do not come from the firm's IT department. Rather, understanding how firms respond to other marketing crises, such as product harm events, provides the best insights for ways to manage a data breach. In Exhibit 6.1, we compare data breaches to other firm crisis events. In particular, marketing crises similarly reduce firm market value [15], lead to abnormal stock returns [16], harm brand equity [17], and create customer churn [18]. The limited research that explores how marketers recover from a product-harm crisis or recall thus can offer important insights for data breaches. For example, some findings indicate that customers and investors interpret firm recovery strategy choices differently; what the market or investors consider a viable, effective approach might simultaneously anger or disappoint customers [19]. Because customer concerns for recovery differ from the market's focus, it is especially important for marketers to have a voice in the firm's overarching recovery strategies, plans, and implementation efforts. However, the scope of and uncertainty surrounding a data breach also mean that easily implementable remedies are elusive for most firms.

An *organizational crisis* involves a situation of escalating intensity that is subject to media scrutiny, legal sanctions, or government intervention, with the potential to jeopardize the firm's reputation or interfere with its business operations, as well as create significant damage to the company's bottom line [20]. Crisis scholars note that these events can threaten both firm reputations and long-term viability [21]. Crisis communication research accordingly can inform data breach recovery best practices, through its detailed focus on response strategies and organizational recovery following a negative firm event. In this research stream, the general approach a company takes toward addressing a crisis has important customer and other performance implications [22]. However, the specific dimensions of a firm's recovery approach can have important ramifications too, including effects on both customer and public responses. For example, adopting particular crisis response strategies are more effective for reducing negative organizational outcomes [23–25] and also can generate heightened feelings of customer forgiveness and reconciliation [26, 27].

Despite the helpful insights that organizational crisis research offers for data breach recovery contexts, the uncertainty of any data breach, in terms of its reach and scope, coupled with firms' limited ability to restore customer well-being fully, make this harm distinct from other firm crises (Exhibit 6.1). Using the framework advanced in Part I of this book, we note that risks to data privacy following a data breach imply heightened customer vulnerability. Any data breach recovery efforts thus must prioritize customer vulnerability mitigation as a focal, immediate goal. Furthermore, customer perceptions of firm benevolence must go hand-in-hand with efforts to reduce vulnerability. These two mechanisms—reducing customer vulnerability and increasing perceptions of firm benevolence—then provide the

Exhibit 6.1 Customer bad news: Types, potential for harm and uncertainty, and firm remedy ability

Customer bad news type	Examples	Threat of harm to customer	Uncertainty regarding harm	Firms' remedy ability	Firm Remedy Examples
Service failure	Sub-par quality; slow delivery; long waits; inadequate performance	Low	Low	High	Apology: price adjustment; future service compensation or discount
Product failure	Damaged goods; product early wear; product malfunction	Low	Low	High	Replacement product; new product price adjustment; repair
Product risk	Product found to be harmful, unhealthy, or somehow high risk in nature, while not necessarily warranting recall or regulatory intervention	Moderate	Low	High	Product warnings, disclaimers, or disclosures; firm may require waivers for use or other legal preventions to avoid sanction
Product crisis	Automobile recall; pharmaceutical product/medical device recall; children's product recall; food contamination	High	Low	High	Product recall informing customers; clearing distribution channels; repair kits/strategies; product replacement
Data breach	Compromise of customer credit card, social security, investment accounts, health details, socially sensitive data, or other personally identifiable information	High	High	Low	Apology: direct messaging from firm or CEO; free credit monitoring offers
Medical health issues	Consumer health diagnosis delivered by physician or other health care provider	High	Varies by diagnosis	Low	Medical or pharmacological treatment program; alternative therapies; comfort; peace of mind

key determinants of whether a company's data breach recovery strategy is effective. We also consider findings and advice from the medical profession on best practices for conveying difficult diagnoses. Together, these insights can establish concrete dimensions for customer-focused recovery strategies (Exhibit 6.1).

> These two mechanisms—reducing customer vulnerability and increasing perceptions of firm benevolence—then provide the key determinants of whether a company's data breach recovery strategy is effective.

Recovery Effectiveness: What Customers Want

Customer Vulnerability and Firm Benevolence

By now, readers should have a strong sense that customer vulnerability is a key theme underlying every new concept we introduce. Mitigating customer vulnerability is always critical to creating beneficial data privacy outcomes [28]. Recall that vulnerability represents customers' perception of their susceptibility to injury or harm [29]. With respect to data privacy issues, customer anxiety becomes most pronounced in situations with a strong potential for damage—even if no data misuse or financial violation occurs [30]. Considering the significant uncertainty and sense of personal risk created by a data breach announcement, we know that customer vulnerability will be high, so unequivocally, reducing customer vulnerability is a necessary component of any firm recovery efforts. If it cannot quell customer vulnerability, the company stands little chance of a full recovery following a data breach. Simply put, for data breach recovery strategies to suppress negative customer behaviors such as reduced purchases or switching, the strategy must reduce feelings of customer vulnerability.

Because data breach recovery is highly firm specific, regardless of fault, we also argue that marketers must work to improve customers' perceptions of the company. *Firm benevolence* captures the extent to which the company is perceived as putting customer needs first, is concerned with customer well-being, and looks out for its customers' best interest [31]. Our empirical examination of data breach recovery outcomes shows that both customer vulnerability and firm benevolence are necessary and effective recovery mechanisms. That is, customers must believe that the company is responding to a data breach in a way that makes them less vulnerable. If the company also appears caring, sympathetic, and benevolent in its recovery efforts, it has a strong positive effect on stabilizing customer relationships. We argue that the various tactical elements of data breach recovery efforts that we introduce work because they make the company seem benevolent, implying that it seeks to protect its customers from harm. Customer vulnerability reduction and firm benevolence are how companies' data breach recovery strategies ultimately can prevent customer defection and preserve strong relationships.

Customer Behavior Effects

Finding the missing links between recovery strategy tactics and how well they work is critical but, thus far, unavailable. These insights may be more timely than ever before. Research shows that following a data breach announcement, customer spending drops by nearly one-third, purchase outings decline by more than 20%, and the number of products purchased drops by over 22%. Although this research also indicates that the effects on negative customer behaviors can be short-lived, their extensiveness depends on the data breach recovery strategy in place [32]. Thus companies must think proactively about their data breach recovery response and establish a readily implementable plan for communicating with and compensating customers.

Customer vulnerability and firm benevolence are key mechanisms directly affected by

> Research shows that following a data breach announcement, customer spending drops by nearly one-third, purchase outings decline by more than 20%, and the number of products purchased drops by over 22%.

recovery strategy elements, but marketers also want to know what their efforts mean for customer behaviors that ultimately will influence firm performance, including future sales and loyalty. We can examine customers' likelihood to continue purchasing from a company using their purchase intentions. We also consider the important notion of customer defection or switching behavior, such that customers move their business from one company to another, rejecting its products and services. Customer vulnerability negatively affects customer performance [33], which likely translates into reduced purchase intentions and increased switching. On the other side, firm benevolence has been linked to beneficial firm outcomes over time [34], and it should promote future purchase intentions and reduce switching. Thus, when we describe data breach recovery efforts to mitigate customer vulnerability and enhance perceptions of firm benevolence, we ultimately are seeking to preserve this portfolio of beneficial customer behavior outcomes. Purchase intentions, future sales, and switching behaviors all feature in our research as intended end goals of any data breach recovery strategy implemented by a firm.

Because we examine data breach recovery strategies from the customer perspective, and consistent with advice from crisis communication researchers and physician training programs, we focus on the responses of those who receive the bad news. Product harm research in marketing often emphasizes company outcomes, but medical literature strongly recommends investigating patient perceptions of a difficult diagnosis and its delivery [35]. Patients interpret the diagnosis and its impact on their health and well-being by inferring meaning from the demeanor of the bad news conveyor—typically, a physician in serious diagnosis situations. We extend this thinking to predict that data breaches and their recovery invoke customer evaluations of their personal harm likelihood (i.e., vulnerability) and of the firm, related to how it manages the breach (i.e., benevolence).

Learning from Bearers of Bad News

The Medical Example

Many medical fields feature patient diagnoses that are uncertain in scope and offer limited options for remediation, not unlike data breaches. In turn, medical literature provides physicians with guidance regarding the best ways to deliver bad news to patients. Clinical oncology and other professions accustomed to conveying serious patient diagnoses offer particularly detailed response strategies, and these prescriptions can suggest implications for business crisis recovery options as well. Specifically, medical training programs encourage physician strategies that exhibit maximum patient and caregiver understanding, reduce uncertainty, and convey empathy when communicating difficult diagnoses [36]. We deconstruct optimal communication strategies according to the delivery method and setting, as well as contents of the communication itself.

For example, the widely cited SPIKES methodology suggests establishing the Setting, Perception, Invitation, Knowledge, Empathy, and Strategy elements to communicate a serious medical diagnosis [37]. Physicians first should consider the timing and context of when and where the diagnosis will be delivered. Some settings simply are not appropriate for or conducive to sharing serious diagnoses. Being sensitive to the patient's state of mind, as well as the presence (or absence) of important others, can suggest the optimal setting for conveying bad news. Physicians also carefully manage the perception they create upon entering the setting. According to SPIKES program evidence, the physician's facial expressions and demeanor can prompt a patient to worry needlessly or formulate opinions before obtaining the facts about the diagnosis. Being mindful of these perceptions of initial demeanors also can increase the odds that the patient will invite the physician into a reciprocal conversation, which is more conducive to sharing a serious diagnosis. By waiting for and securing such an invitation from the patient, the physician also creates a more level playing field between the two.

The "k" element means that the physician shares everything known, as well as specifies what remains unknown, in a clear way. Sharing knowledge carefully, so that the patient understands the situation accurately, is important. It may require explaining complex topics more than once and in more than one way, leaving ample opportunities for questions and follow-up discussion. Depending on the diagnosis, patients might struggle to process highly detailed or complicated information, but it is up to the physician to establish the patient's understanding. This goal relates to the strategy element, in that physicians need a clear, step-by-step response strategy ready to share with the patient. As soon as a patient processes the elements of the diagnosis, she or he can begin to buy into proposed recovery or harm mitigation strategies. Finally, the conveyor of bad news can increase understanding and create a greater sense of ease by exhibiting a strong sense of empathy. Signaling to the patient that the physician will be working together with the patient, and realizes how difficult coping with such news can be, is an important component of the overall delivery method. Medical diagnoses often are far more serious than news of a data breach, yet their important similarities make this setting instructive for determining ways to convey information about breaches too (Exhibit 6.2).

Organizational crisis research also evaluates the parameters for delivering bad news, in a business setting [38]. They closely parallel medical literature and its SPIKES framework. The foundations for this research stream include bad news delivery timing and medium, face management and self-presentation, account giving, truth-telling, and information disclosure. By thinking through how, who, what, and where questions before conveying bad news, firms in

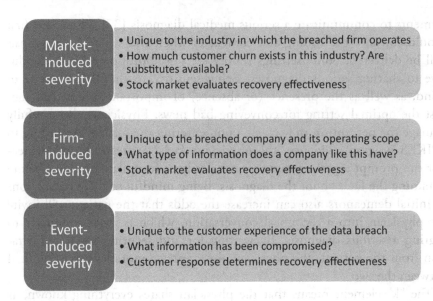

Market-induced severity
- Unique to the industry in which the breached firm operates
- How much customer churn exists in this industry? Are substitutes available?
- Stock market evaluates recovery effectiveness

Firm-induced severity
- Unique to the breached company and its operating scope
- What type of information does a company like this have?
- Stock market evaluates recovery effectiveness

Event-induced severity
- Unique to the customer experience of the data breach
- What information has been compromised?
- Customer response determines recovery effectiveness

Exhibit 6.2   How bad is it? Sources of Data Breach Severity

general, and marketers in particular, may be able to create and deliver crisis communication

> By thinking through how, who, what, and where questions before conveying bad news and considering its likely interpretation by recipients, firms in general, and marketers in particular, can create and deliver crisis communication messages that are better understood and evaluated by customers.

messages that will be better understood and evaluated by customers. Their initial contacts then can set the stage for successful recovery efforts in the wake of a data breach.

How Bad Is It? Sources of Severity

Not all data breach events are created equal. Some events may affect a few hundred or a thousand customers (e.g., Fuzzy's Taco Shop, affecting 2000 customers); others are truly massive in scope (e.g., FriendFinder.com, affecting 412 million customers). Data breach severity also may differ with the type of information collected and exposed, which varies across retail, financial services, and healthcare categories, for example. Crisis communication perspectives argue that crisis events vary in intensity and severity [39]. We concur with and extend this thinking; the severity of a data breach should determine the amount of customer vulnerability too. Thus, to achieve a more fine-grained understanding, we consider the severity generated from various sources relevant to the breach: (1) market-induced severity, (2) firm-induced severity, and (3) event-induced severity related to the actual breach (Exhibit 6.2). Market- and firm-induced severity may be most germane to investor expectations of harm; event-induced severity is most relevant to customer expectations about data breach harms. With this diverse approach, companies can appreciate the full picture of likely harm associated with both market evaluations and customer responses.

Market-induced severity exists when turbulent and volatile industries create information asymmetry and overall market instability, including heightened customer churn [40]. With greater information asymmetry and market instability, investors take particularly careful note of the recovery strategy delivery elements following a data breach. The way the data breach recovery announcement is communicated and by whom both send powerful signals to the market about the anticipated damage, which could induce even greater customer churn. For example, having the CEO express the data breach recovery communication may signal to investors that considerable harm is possible, causing them to revise their performance expectations downward. The high switching rates and customer volatility in these markets should increase even further with data breach recovery announcements [41]. That is, high market-induced severity will worsen the negative firm performance effects resulting from market and investor expectations.

Firm-induced severity also pertains to the market, in that it accounts for the specific form of severity created by the breached firm's business operations. Evidence of firm-induced severity arises from event studies that examine the impact of data breaches on abnormal stock returns and that distinguish the variation that arises in sensitive firm contexts, such as health care and financial services [42]. Product recall research similarly predicts negative investor expectations for firms operating in more sensitive categories, such as those that make products for children [43, 44]. Big data consultants hold that consumers are willing to pay more to protect the personal information they have provided to medical and financial services firms, relative to that held by firms that collect less sensitive personal information [45]. Collectively, this evidence implies heightened investor expectations of performance detriments, related to customer backlash and switching, when firm-induced severity is greater. Such firm-induced severity may best be understood according to various data breach recovery strategy content elements. Because this form of severity stems from the firm, the firm must offer restorative measures to improve customers', and ultimately the market's, perceptions of its recovery. Firm apologies and compensation offer two such measures.

Finally, a third form of severity results from the event itself and is perceived by the customer, rather than the market or investors (as occurs for market- and firm-induced severity). *Event-induced severity* is unique to each customer's experience of the data breach. It encapsulates a customer's perceptions of what information has been compromised and personal assessments of the likelihood that she or he actually will be the victim of fraudulent activity or identity theft. Event-induced severity likely will color all customer evaluations of a data breach recovery, pertaining to both the delivery and content elements of these recovery strategies. In the discussion that follows, we disentangle various approaches available to firms to communicate a data breach. We specifically consider the content of the recovery strategy and the ways it gets delivered. These ideas constitute strategic guidance for marketers to put their firm on a path to reconnecting effectively with customers and restoring relationships (Exhibit 6.2).

Customer-Focused Data Breach Recovery Strategies

Crisis communication theory offers a simple, elegant conceptualization of strategies for conveying serious bad news, grouped by their instructive and accommodative elements [46]. In practice, these categories manifest as recovery strategy *delivery* elements (instructive) and *content* elements (accommodative). Put simply, they represent the "how" and the "what" of crisis communication. We apply this framework to a data breach recovery setting, as one form of organizational crisis and a serious form of customer bad news. Two recovery strategy delivery approaches may inform customers of the data breach recovery details: (1) *customer-direct delivery* of pertinent recovery information and (2) *authoritative delivery* by an important firm representative, such as someone from the executive or C-suite. Furthermore, different content strategies may be designed to comfort or assist customers following a data breach event, in the form of (1) a *firm apology* or (2) provisions of *protective compensation*. We describe each recovery strategy element and its effects on customer vulnerability and firm benevolence in greater detail next.

Recovery Strategy Delivery

Customer-Direct Delivery. Crisis responses using a customer-direct communication approach represent a voluntary method by the firm to message its customers about a data breach. In contrast, customers might learn about a data breach through the media, an indirect source such as a company website or blog, or some other indirect communication channel. Academic research in marketing suggests that voluntary response strategies, or customer-direct approaches, are better received in product crisis events [47]. Telling customers honestly and directly about a negative event can signal that the company has contained the damage and has the crisis situation under control [48]. Direct messaging also encourages customer perceptions of reduced vulnerability and minimizes the potential for future harm [49].

> Telling customers honestly and directly about a negative event can signal that the company has contained the damage and has the crisis situation under control. Direct messaging also encourages customer perceptions of reduced vulnerability and minimizes the potential for future harm.

According to crisis communication research, more forthright firm communication signals a smaller likelihood of damage due to the crisis event. Beyond perceptions of damage and harm, a company's willingness to communicate directly establishes a sense that the company cares, is emotionally supportive [50], and values its customer relationships [51]. Finally, in reviews of organizational bad news research, we find empirical support for the notion that richer media communication (i.e., customer-direct vs. indirect messages) in a crisis seems more benevolent and generally is more favorably evaluated by the intended recipient [52].

Despite this evidence and its intuitive appeal, many firms avoid communicating with customers directly following a data breach. Examples such as the eBay data breach story we told earlier in this chapter show that companies often elect to use far less evident messaging techniques. Their reasons may include concerns about future litigation or an insufficient understanding of the breach details. But regardless of these motivations, a

customer-direct recovery strategy, rather than indirect customer communication via a website, blog post, or the media, will be better received by customers, because it can suppress customer vulnerability and make the firm
appear more benevolent. Overall, an honest, direct delivery approach is a
better data breach recovery strategy, and better for overall firm performance.

Authoritative Delivery. Authoritative recovery strategies imply that the crisis communication with customers is delivered by the CEO or some other
influential representative

> Authoritative recovery strategies in which the C-suite issues messages to cus-
> tomers can serve as signals of extreme data breach harm and make customers
> worry unnecessarily.

of the C-suite. Many of the same arguments relevant to customer-direct
delivery apply to the delivery of data breach information to customers by
someone in authority. In particular, crisis communications from an important member of the firm can be a sign of a valued relational approach and
firm benevolence [53]. Customers sense that the firm actually regards their
relationship as important, even at the highest levels. Similarly, ethical analyses that deconstruct firm crisis responses indicate that having the CEO
deliver the message is both helpful and recommended, as a component of
efforts to make amends to customers [54].

However, involving a member of the C-suite in communications about
damage also might worsen perceptions of vulnerability. Empirical evidence
suggests that messages delivered by authority figures heighten perceptions of
event damage and harm [55]. The idea is that only extensive threats of customer harm or a strong probability of damage could prompt the involvement of such high-level decision makers. Although the precise role of the
CEO in product harm crises and recovery has received limited attention, we
find that the negative implications of authoritative recovery strategies can
overwhelm their positive effects, by increasing event salience and perceptions
of vulnerability. Medical bad news literature confirms these notions; patients
interpret health communications and diagnoses as more severe if they come
from figures with greater authority [56]. Authoritative recovery strategies
involving the C-suite thus might serve as a signal of extreme data breach
harm that causes customers to worry, perhaps more than is justified by the
actual harm. In this case, perceptions of firm benevolence likely get over-
whelmed by heightened feelings of risk.

Recovery Strategy Content

Firm Apology. Some crisis communication experts suggest that apologies are risky recovery strategies, with the potential to make a bad situation worse by encouraging blame attributions [57]. An apology signals company guilt, so it makes matters worse. Other experts also advocate against apologizing, citing evidence that company apologies invite litigation [58]. The vast body of academic research on apologies actually shows mixed effects for remedying a crisis though [59]. In our analysis of these research findings, we realized that this research stream often examines only whether apologies can dampen consumer anger. Although we agree that customers are likely to be angry following a data breach, we propose that other customer emotions and cognitions are worth considering too when it comes to preserving the relationship with the firm.

In this sense, crisis recovery literature often emphasizes the value of an apology, above and beyond other recovery strategy content [60]. Service failure recovery research in the marketing

> Citing evidence that an apology, empathy, or atonement can restore customer relationships, we assert that a firm apology following a data breach will be well-received by customers, markets, and investors, so it should lessen negative performance effects.

field identifies an apology as a powerful, benevolent signal that the firm wants to make amends for customer harms [61]. An apology can restore and even improve customer satisfaction following a service failure [62]. Crisis recovery perspectives also cast apologies as both necessary and beneficial steps toward repairing the customer–firm relationship [63]. Similarly, medical literature cites the more positive patient evaluations and interpretations that result when the physician shows empathy for the recipient, acknowledging the bad news they are being forced to receive [64]. Even if apologies create perceptions of firm benevolence though, they are unlikely to reduce customer vulnerability. Nonetheless, considering substantial evidence that an apology, empathy, or atonement can restore customer relationships, we assert that a firm apology following a data breach will be well-received by customers, markets, and investors, so it should lessen negative performance effects. Our own research on data breach recovery supports the idea that saying "sorry" is a difficult but necessary first step in reconnecting with customers.

Protective Compensation. Crisis communication theory describes a class of recovery strategy content that is aimed at correcting the balance between the customer and the firm through an offer of an equity exchange. In a data breach context, firms often use protective compensation recovery strategies, such as providing free credit monitoring services for customers. Credit monitoring, by its very nature, suggests that the customer faces some threat of harm and must take control to manage and mitigate this damage. Therefore, these offers may trigger customer feelings of anxiety and risk and, ultimately, perceptions of greater threat than actually exists. The unintended consequence may be to increase feelings of vulnerability. Empirical studies in the apology domain also show that when a firm's compensation offer is not fully restorative, as is true of free credit monitoring, the strategy may do more harm than good [65]. Credit protection services only identify when a customer's personal information has been used fraudulently (e.g., unauthorized accounts opened with the customer's name); they cannot prevent this fraudulent activity, nor can they retrieve and re-secure compromised information. In that sense, this recovery strategy fails to be fully restorative.

However, a data breach is unique, relative to other product harm crises that have identifiable solutions or fixes. In this case, protective compensation through credit protection services may represent an important and viable option for firms attempting to restore the equity exchange and provide at least some means to reduce vulnerability. Because a data breach represents a unique type of customer harm, there is some reason to expect that a protective compensation strategy could create beneficial effects in this particular setting. We draw on evidence in product recall marketing literature [66], which indicates that offering customers a "full remedy" is better received than a partial remedy. However, full remedies (cf. partial remedies) tend to be evaluated less favorably by investors and the market, because a full remedy signals the extent of customer harm and damage. In own research, we find that a protective compensation recovery strategy (i.e., free credit monitoring) signals significant customer harm and breach damage to investors, with potentially negative effects on firm stock market performance—even if this strategy also improves customer effects. Still, in modern environments, where companies and customers frequently are plagued by data breaches, offering free credit protection seemingly has become an expected minimum reparation for customers. It signals firm benevolence and provides some form of vulnerability mitigation, even if an insufficient one. Thus, we recommend such reparation undertaken in a broader context of making customer amends, with the understanding that these efforts may have unintended market outcomes.

Recovery Strategy	Alternative Approaches	Benefits	Drawbacks	Vulnerability Reduction	Sign of Firm Benevolence
Customer-Direct Delivery	Media release; company posts on website; no public announcement	Direct communication expresses customer concern; ensures they receive message	Customers may become more worried than necessary in small breaches	High	Moderate
Authoritative Delivery	Information comes from employee not in C-suite; anonymous source (e.g., Data Security Team)	Lets customers know even highest levels are managing situation; shows the "top" cares	Can signal the breach is substantial, if it requires the CEO to be involved	Moderate	High
Company Apology	Attributing blame to hackers, other external source; not taking responsibility	Shows the company cares and sends a strong signal of empathy	Attribution of blame can invite litigation; signal to the market that greater company losses are expected	Low	High
Protective Compensation	Information provision only; financial compensation such as discounts; providing nothing	Customers now expect this service; directly compensates for harm	May be overkill for small breaches; programs can be costly, signaling likely losses to the stock market	High	Moderate

Exhibit 6.3 Recovery strategies: Ability to reduce vulnerability and show benevolence

We detail the different recovery strategies and their effects on customer vulnerability and perceptions of firm benevolence in Exhibit 6.3.

Keeping the Customer Focal: One More Reminder

As another reminder: We advise marketing teams to craft their recovery strategies in ways that explicitly consider customers' centrality. It is their personal information that has been deemed valuable enough (and vulnerable enough) for hackers to pursue. The company's immediate instinct in reaction to a data breach event may be to avoid harm to the firm, but it should be to consider the individual customers whose personal, private, and possibly sensitive information is now in the hands of criminals. The frequency with which data breaches occur

In modern environments, where companies and customers frequently are plagued by data breaches, offering free credit protection seemingly has become an expected minimum reparation for customers. It signals firm benevolence and provides some form of vulnerability mitigation, even if an insufficient one.

make them no less scary to the people who must now endure their consequences. Among victims of data breaches, nearly half sense that their identity is at risk "forever" [67].

Data breaches thus have become an unfortunate but regular occurrence in today's business environment. The preceding discussion suggests a portfolio of recovery strategies that have demonstrated ability to reduce customer vulnerability and promote firm benevolence. These elements function as links between how a company responds to a data breach and the customer reactions and behaviors that such a breach can invoke. Reducing vulnerability and promoting benevolence can keep customers from defecting to a competitor. And though extant evidence reveals a drop in sales immediately following a data breach, these recommended mechanisms can work to stabilize the long-term effects on future sales, purchase intentions, and goodwill toward the firm. As Adam Levin wisely mused, "Get something wrong in the realm of corporate communications, and you can count on hearing about it. Even subtle public relations missteps can do irreparable harm to your brand. Get something right, and most of the time no one notices" [68]. In spite of this truism, we believe this chapter offers sound advice for managers to recover from the harms of a data breach.

References

1. For More Information Consult the Privacy Rights Clearinghouse. *Learn About Privacy*. www.privacyrights.org.
2. See www.idtheftcenter.org for More Information.
3. Rouse, M. (2017, December). *What Is a Data Breach?* TechTarget. Available at https://searchsecurity.techtarget.com/definition/data-breach. Accessed July 23, 2018.
4. Faulkner, B. (2007). Hacking into Data Breach Notification Laws. *Florida Law Review, 59*, 1097–1125.
5. For Further Information see the *National Conference of State Legislatures*. (2018). 2018 Security Breach Legislation. Updated May 6, 2018. Available at http://www.ncsl.org/research/telecommunications-and-information-technology/2018-security-breach-legislation.aspx. Accessed July 23, 2018.
6. Martin, K. D., Borah A., & Palmatier, R. W. (2018, February 15). A Strong Privacy Policy Can Save Your Company Millions. *Harvard Business Review*. www.hbr.org.
7. Smith, H. J., Dinev, T., & Xu, H. (2011, December). Information Privacy Research: An Interdisciplinary Review. *MIS Quarterly, 35*, 980–1035.
8. Romanosky, S., Hoffman, D., & Acquisti, A. (2014). Empirical Analysis of Data Breach Litigation. *Journal of Empirical Legal Studies, 11*(1), 74–104.
9. Coombs, W. T. (2007). Attribution Theory as a Guide for Post-crisis Communication Research. *Public Relations Review, 33*, 135–139.
10. Schwartz, P. M., & Janger, E. J. (2007). Notification of Data Security Breaches. *Michigan Law Review, 105*, 913–985.

11. Burg, N. (2014, January 17). Five Lessons for Every Business from Target's Data Breach. *Fortune*. Available at www.forbes.com. Accessed October 24, 2016.

12. Yang, J. L., & Amrita Jayakumar. (2014, January 10). Target Says Up to 70 Million More Customers Were Hit by December Data Breach. *Washington Post*. Available at www.washingtonpost.com/business/economy. Accessed October 25, 2016.

13. Greenberg, A. (2014, May 23). eBay Demonstrates How Not to Respond to a Huge Data Breach. *Wired*. Available at www.wired.com. Accessed October 25, 2016.

14. See https://krebsonsecurity.com/tag/michaels-breach/.

15. Gao, H., Xie, J., Wang, Q., Wilbur, K. C. (2015, September). Should Ad Spending Increase or Decrease Before a Recall Announcement? The Marketing-Finance Interface in Product-Harm Crisis Management. *Journal of Marketing, 79*, 80–99.

16. Germann, F., Grewal, R., Ross, W. T., Jr., & Srivastava, R. K. (2014). Product Recalls and the Moderating Role of Brand Commitment. *Marketing Letters, 25*, 179–191.

17. Dawar, N., & Pillutla, M. M. (2000, May). Impact of Product-Harm Crises on Brand Equity: The Moderating Role of Consumer Expectations. *Journal of Marketing Research, 37*, 215–226.

18. Knox, G., & van Oest, R. (2014, September). Customer Complaints and Recovery Effectiveness: A Customer Base Approach. *Journal of Marketing, 78*, 42–57.

19. Liu, A. X., Liu, Y., & Luo, T. (2016, May). What Drives a Firm's Choice of Product Recall Remedy? The Impact of Remedy Cost, Product Hazard, and the CEO. *Journal of Marketing, 80*, 79–95.

20. Fink, S. (1986, March). Crisis Forecasting. *Management Review, 75*, 52–57.

21. Pearson, C. M. (1993). From Crisis Prone to Crisis Prepared: A Framework for Crisis Management. *Executive, 7*, 48–59.

22. Coombs, W. T. (2007). Attribution Theory as a Guide for Post-crisis Communication Research. *Public Relations Review, 33*, 135–139.

23. Coombs, W. T., & Holladay, S. J. (2008). Comparing Apology to Equivalent Crisis Response Strategies: Clarifying Apology's Role and Value in Crisis Communication. *Public Relations Review, 34*, 252–257.

24. Fehr, R., & Gelfand, M. J. (2010). When Apologies Work: How Matching Apology Components to Victims' Self-Construal Facilitates Forgiveness. *Organizational Behavior and Human Decision Processes, 113*, 37–50.

25. Lee, S., & Chung, S. (2012, December). Corporate Apology and Crisis Communication: The Effect of Responsibility Admittance and Sympathetic Expression on Public's Anger Relief. *Public Relations Review, 38*, 932–934.

26. Kirchhoff, J., Wagner, U., & Strack, M. (2012). Apologies: Words of Magic? The Role of Verbal Components, Anger Reduction, and Offense Severity. *Peace and Conflict: Journal of Peace Psychology, 18*(2), 109–130.

27. Koehn, D. (2013, April). Why Saying 'I'm Sorry' Isn't Good Enough: The Ethics of Corporate Apologies. *Business Ethics Quarterly, 23*, 239–268.

28. Martin, K. D., Borah A., & Palmatier, R. W. (2017, January). Data Privacy: Effects on Customer and Firm Performance. *Journal of Marketing, 81*, 36–85.

29. Smith, N. C., & Cooper-Martin, E. (1997, July). Ethics and Target Marketing: The Role of Product Harm and Consumer Vulnerability. *Journal of Marketing, 61*, 1–20.

30. Fisher, J. A. (2013). Secure My Data or Pay the Price: Consumer Remedy for the Negligent Enablement of Data Breach. *William & Mary Business Law Review, 215*(4), 217–233.

31. Victor, B., & Cullen, J. B. (1988, March). The Organizational Bases of Ethical Work Climates. *Administrative Science Quarterly, 33*, 101–125.

32. Janakiraman, R., Lim, J. H., & Rishika, R. (2018, March). The Effect of a Data Breach Announcement on Customer Behavior: Evidence from a Multichannel Retailer. *Journal of Marketing, 82*, 85–105.

33. Martin, K. D., Borah, A., & Palmatier, R. W. (2017, January). Data Privacy: Effects on Customer and Firm Performance. *Journal of Marketing, 81*, 36–85.

34. Martin, K. D., & Cullen, J. B. (2006, December). Continuities and Extensions of Ethical Climate Theory: A Meta-Analytic Review. *Journal of Business Ethics, 69*, 175–194.

35. Alby, F., Zucchermaglio, C., & Fatigante, M. (2016, July 14). Communicating Uncertain News in Cancer Consultations. *Journal of Cancer Education* (published electronically). https://doi.org/10.1007/s13187-016-1070-x.

36. Ptacek, J. T., & Eberhardt, T. L. (1996). Breaking Bad News: A Review of the Literature. *Journal of the American Medical Association, 276*, 496–502.

37. Buckman, R. A. (2005, March/April). Breaking Bad News: The S-P-I-K-E-S Strategy. *Community Oncology, 2*, 138–142.

38. For a Comprehensive Review, see Bies, R. J. (2013, January). The Delivery of Bad News in Organizations: A Framework for Analysis. *Journal of Management, 39*, 136–62.

39. Stephens, K. K., Malone, P. C., & Bailey, C. M. (2005, October). Communicating with Stakeholders During a Crisis. *Journal of Business Communication, 42*, 390–419.

40. Fang, E., Er., Palmatier, R. W., & Steenkamp, Jan-Benedict E. M. (2008, September). Effect of Service Transition Strategies on Firm Value. *Journal of Marketing, 72*, 1–14.

41. Knox, G., & van Oest, R. (2014, September). Customer Complaints and Recovery Effectiveness: A Customer Base Approach. *Journal of Marketing, 78*, 42–57.

42. Malhotra, A., & Malhotra, C. K. (2011). Evaluating Customer Information Breaches as Service Failures: An Event Study Approach. *Journal of Service Research, 14*(1), 44–59.

43. Chen, Y., Ganesan S., & Liu, Y. (2009, November). Does a Firm's Product-Recall Strategy Affect Its Financial Value? An Examination of Strategic Alternatives During Product Harm Crises. *Journal of Marketing, 73*, 214–226.

44. Liu, A. X., Liu, Y., & Luo, T. (2016, May). What Drives a Firm's Choice of Product Recall Remedy? The Impact of Remedy Cost, Product Hazard, and the CEO. *Journal of Marketing, 80*, 79–95.

45. Morey, T., Forbath, T. "Theo", & Schoop, A. (2015, May). Customer Data: Designing for Transparency and Trust. *Harvard Business Review, 93*, 96–105.

46. Coombs, W. T. (2007). Attribution Theory as a Guide for Post-crisis Communication Research. *Public Relations Review, 33*, 135–139.

47. Dawar, N., & Pillutla, M. M. (2000, May). Impact of Product-Harm Crises on Brand Equity: The Moderating Role of Consumer Expectations. *Journal of Marketing Research, 37*, 215–226.

48. Gao, H., Xie, J., Wang, Q., & Wilbur, K. C. (2015, September). Should Ad Spending Increase or Decrease Before a Recall Announcement? The Marketing-Finance Interface in Product-Harm Crisis Management. *Journal of Marketing, 79*, 80–99.

49. Chen, Y., Ganesan, S., & Liu, Y. (2009, November). Does a Firm's Product-Recall Strategy Affect Its Financial Value? An Examination of Strategic Alternatives During Product Harm Crises. *Journal of Marketing, 73*, 214–226.

50. Weidan, C., Qi, X., Yao, T., Han, X., & Feng, X. (2016). How Doctors Communicate the Initial Diagnosis of Cancer Matters: Cancer Disclosures and its Relationship with Patients' Hope and Trust. *Psycho-Oncology* (published electronically). https://doi.org/10.1002/pon.4063.

51. Ringberg, T., Odenkerken-Schroder, G., & Christensen, G. L. (2007, July). A Cultural Models Approach to Service Recovery. *Journal of Marketing, 71*, 194–214.

52. Bies, R. J. (2013, January). The Delivery of Bad News in Organizations: A Framework for Analysis. *Journal of Management, 39*, 136–162.

53. Liu, A. X., Liu, Y., & Luo, T. (2016, May). What Drives a Firm's Choice of Product Recall Remedy? The Impact of Remedy Cost, Product Hazard, and the CEO. *Journal of Marketing, 80*, 79–95.

54. Koehn, D. (2013, April). Why Saying 'I'm Sorry' Isn't Good Enough: The Ethics of Corporate Apologies. *Business Ethics Quarterly, 23*, 239–268.

55. Bies, R. J. (2013, January). The Delivery of Bad News in Organizations: A Framework for Analysis. *Journal of Management, 39*, 136–162.

56. See Ptacek, J. T., & Eberhardt, T. L. (1996). Breaking Bad News: A Review of the Literature. *Journal of the American Medical Association, 276*, 496–502.

57. Kim, P. H., Ferrin, D. L., Cooper, C. D., & Dirks, K. T. (2004). Removing the Shadow of Suspicion: The Effects of Apology Versus Denial for Repairing Competence- Versus Integrity-Based Trust Violation. *Journal of Applied Psychology, 89*(1), 104–118.

58. Coombs, W. T., & Holladay, S. J. (2008). Comparing Apology to Equivalent Crisis Response Strategies: Clarifying Apology's Role and Value in Crisis Communication. *Public Relations Review, 34,* 252–257.
59. See Skarlicki, D. P., Folger, R., & Gee, J. (2004, February). When Social Accounts Backfire: The Exacerbating Effects of a Polite Message or an Apology on Reactions to an Unfair Outcome. *Journal of Applied Social Psychology, 34,* 322–341.
60. Benoit, W. L., & Drew, S. (1997). Appropriateness and Effectiveness of Image Repair Strategies. *Communication Reports, 10*(Summer), 153–163.
61. Torsten, R., Odenkerken-Schroder, G., & Christensen, G. L. (2007, July). A Cultural Models Approach to Service Recovery. *Journal of Marketing, 71,* 194–214.
62. Wirtz, J., & Mattila, A. S. (2004). Consumer Responses to Compensation, Speed of Recovery and Apology After a Service Failure. *International Journal of Service Industry Management, 15*(2), 150–166.
63. Koehn, D. (2013, April). Why Saying 'I'm Sorry' Isn't Good Enough: The Ethics of Corporate Apologies. *Business Ethics Quarterly, 23,* 239–268.
64. Fallowfield, L., & Jenkins, V. (2004, January). Communicating Sad, Bad, and Difficult News in Medicine. *Lancet, 363,* 312–319.
65. Fehr, R., & Gelfand, M. J. (2010). When Apologies Work: How Matching Apology Components to Victims' Self-Construal Facilitates Forgiveness. *Organizational Behavior and Human Decision Processes, 113,* 37–50.
66. See Liu, A. X., Liu,Y., & Luo, T. (2016, May). What Drives a Firm's Choice of Product Recall Remedy? The Impact of Remedy Cost, Product Hazard, and the CEO. *Journal of Marketing, 80,* 79–95.
67. Ponemon Institute. (2014). The Aftermath of a Data Breach: Consumer Sentiment. *Ponemon Institute Research Report.* www.ponemon.org.
68. Levin, A. (2018, June 7). This Company Was the Latest to Suffer a Data Breach. Its Reaction Was Perfect. *Inc.* Available at https://www.inc.com/adam-levin/this-company-was-latest-to-suffer-a-data-breach-its-reaction-was-perfect.html.

Part III

Offensive Strategies: Competing with Privacy

7

Understanding and Valuing Customer Data

Introduction

Firms such as Google and Amazon tout the value they generate from the skyrocketing amount of customer data they possess, which support product innovation efforts, tailored offerings, and targeted advertising. The revenues that can be earned by leveraging big data and customer data analytics also are projected to grow more than 50% over the five years from 2015 to 2019, reaching $187 billion worldwide [1]. These sources of revenue reflect a new approach, such that "The capture of such data lies at the heart of the business models of the most successful technology firms (and increasingly, in traditional industries like retail, health care, entertainment and media, finance, and insurance) and government assumptions about citizens' relationship to the state," according to Leah Lievrouw, Professor of Information Studies at the University of California-Los Angeles [2].

As valuable as customer data may be though, they also create perceived risk and vulnerability among customers, due to the potential for data breaches, identity theft, unwanted usage of their personal data, and general feelings of anxiety and vulnerability. Consumers thus tend to express negative reactions to any data collection efforts and a broad unwillingness to share personal data. This trend is especially notable in the recent movement to "delete Facebook," in response to the data breach incident associated with Cambridge Analytica [3]. As one user explained, "I suspected this stuff was

The analyses and writing for this chapter was done with assistance from Jisu Kim, a doctoral student at University of Washington.

going on, but this is the first time it's been plainly exposed…It seems so malicious, and Facebook seems so complicit all the way up and down, like it doesn't care about its users." Similar feelings of violation due to the data breach have led some consumers to switch to Twitter and Instagram to meet their social media needs [3]. Such examples of infuriated customers and their responses to issues are visible in social media posts, but they also affect stock prices [4]. In turn, the Gartner consultancy predicts that spending on information security will reach $93 billion in 2018 [5].

Notably then, even as firms increasingly depend on the monetization of customer data, customers seek to move in an opposite direction and want more privacy. In other words, firms seek to capture and monetize more customer data; customers want to minimize data sharing and avoid firms that might use their data in ways that are contrary to their preferences. When it comes to privacy, the gap between firms' perspective and customers' is growing. It is up to managers to take a proactive stance to ensure that they can generate value for their firms by motivating customers to remain engaged by granting them sufficient value as well. To do so, managers may need to reposition their data management strategies, with innovative options such as offering differential pricing based on data sharing levels or requiring paid subscriptions for previously free services, as Mark Zuckerberg acknowledged in his Senate testimony following Facebook's data privacy scandal [6]. In particular, firms need to find a way to deliver but also communicate about the value that customers may receive from sharing their data. Without such insights, customers have no reason not to rebel, such as by actively seeking an alternative provider with better privacy standards and data policies that they perceive as more fair. Alternatively, future legislation might restrict data practices that currently are considered acceptable or expand the enforcement of the European General Data Protection Regulation (GDPR) requirements worldwide (see Chapter 3), as politicians work to respond to voters' anger about questionable data activities.

To engage in more effective strategizing in relation to customer data privacy, managers also must understand explicitly how customers value their personal data. How much are personal data worth, from customers' perspective? Developing an answer to this question can establish significant implications for both research and practice, because it bridges the gap between customers' expectations and firms' efforts, while also helping firms strategically collect and manage the vast potential of big data in a way that customers perceive as fair. In one study, despite their growing data privacy concerns, among 607 U.S. consumers, 55% of them still indicated they would be willing to share their data in exchange for free services [7]. By recognizing and understanding this form of psychological ambivalence, firms can determine the actual value of personal data for customers, then deliver valuable offerings that account for that information. With this chapter, we focus on four key areas regarding the value of customer data, by

1. Describing the firm's perspective on the value of customer data.
2. Describing the customer's perspective on the value of their data.
3. Reporting on the results of a study, undertaken for this book to specify how customer data value varies across different contexts and types of customers.
4. Proposing a strategy for proactively sharing the value of collected with customers, so that the firm's data policies become a positive feature in its marketing strategy.

> To engage in more effective strategizing in relation to customer data privacy, managers also must understand explicitly how customers value their personal data.

Firm Perspectives on Data Value

Sources of Firm Value

It may seem intuitive to anticipate that companies want to monetize customer data, yet the amount of value that such data provide to companies, and the strategic channels through which that value moves, remain unclear. Key sources of revenue generation that rely on customer data, in order of their increasing potential for creating value but also their risks of imposing customer data vulnerability (Exhibit 7.1), include product improvements, targeted advertising, and resale of the information to third parties. That is, products and services should be enhanced to satisfy customers, according to insights gathered from vast data; the firm then can better target customers according to their heterogeneous needs. Finally, firms can sell these customer data to non-competitive third parties, as long as they do not violate customers' trust in ways that drive the customers away.

Product Improvements. Firms have long used various types of customer data to inform their efforts to enhance existing product and service features. Yet this revenue source remains nontrivial, because of the scale of its

Exhibit 7.1 Examples of data use levels

Data use level	Example	Consumer benefits	Consumer vulnerability level
Product improvement	Amazon's personalized recommendations; Netflix's and Spotify's curated product offerings based on previous activity; Airbnb's search results generated on the basis of prior customer-host interactions, demographics, and search history	High	Low
Targeted advertising	Google and Facebook ads that reflect users' browsing history and demographics	Medium	Low to medium
Resale to third party	UnitedHealth reselling its claim forms data to drug companies; Toyota reselling its GPS traffic data to municipal planning departments and corporate delivery fleets	Low/None	High

potential impact on a company's entire business model. Amazon's recommendation system is a quintessential example: It leverages customers' search activity data to earn $12.83 billion in sales [8]. Most Airbnb hosts might seek the highest prices possible, but Airbnb's Price Tips feature, which assesses people's search terms, timing, and neighborhood preferences, gives hosts more information about the acceptable price range that will maximize purchase likelihood. When the price they list is within 5% of the suggested price, hosts are nearly four times as likely to receive a booking [9].

Targeted Advertising. Their platform business models grant them access to a wide variety of behavioral data, such as searches, social interactions, views, clicks, and purchases, so Google and Facebook account for more than 58% of the digital advertising market [10] (Exhibit 7.2). A report by McKinsey Global Institute further states that companies that use customer data in general outperform their competitors by 85% when it comes to sales growth and more than 25% in gross margins [11]. Thus a worldwide survey of more than 700 organizations reveals that their efforts—to increase spending on customer data analytics to enhance their competitive intelligence about future markets, improve customer targeting, and optimize operations—have generated profit increases of approximately 6%. As these data suggest, more than just the top few big players reap the benefits from customer data.

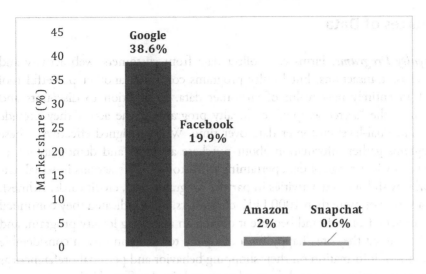

Exhibit 7.2 Digital advertising market share graphic (2017) (*Source of data* Liberto 2018; eMarketer)

Selling Data to Third Parties. Companies also can sell data to third parties that are not their competitors, often operating in a different industry. For example, UnitedHealth accrued $5 billion in revenues by reselling aggregate information it has collected from insurance claims to drug companies. These companies purchase the data from UnitedHealth to analyze consumers' drug usage rates, efficiency, and overall performance compared with competitors' [12]. Toyota also sold its traffic data, generated from GPS navigation devices installed in cars, to municipal planning departments and corporate delivery fleets. Such data sales seem like win-win solutions, because both parties benefit, and the outcomes also help consumers. In other cases, though, selling customers' data might alienate them; one large China telecom provider sold its customers' geolocation data to banks, which sought to identify whether potential clients were good risks for loans, based on how much time they spent at casinos in Macau. In this case, most customers likely perceived that the telecom provider had violated their trust.

Some consultants anticipate large, untapped potential in the data resale domain; only about 5% of all the digital data in the world currently are being leveraged to support analytics efforts [12]. The entire digital universe will likely reach 163 zettabytes by 2025 [13], and 1 zettabyte is equivalent to 1,000,000,000,000,000,000,000 characters, suggesting that there are substantial data still to be analyzed.

Sources of Data

Loyalty Programs. Firms can collect data from customers' web activity and purchase transactions, but loyalty programs constitute another powerful tool and an entirely new realm of customer data. In addition to discounts and rewards, the "secret weapon" of loyalty programs is the access they provide to individual-level customer data over time. When designed effectively, these programs gather information about purchase activities and demographics, as well as a wider range of data pertaining to customers' interactions in social networks, wish lists, and activities in partner programs (e.g., credit cards, airlines). In a survey of more than 1000 U.K. consumers, 87% indicated they continued to purchase from a brand because it offered an appealing loyalty program, and 64% claimed they are "happy for a company to retain and use a considerable amount of information on their shopping behavior and personal preferences as long as it continues to send them relevant and timely offers" [14].

> Loyalty programs may seem like just a platform for discounts and rewards, but their "secret weapon" is the access they provide to *individual-level customer data over time.*

In its effort to determine which rewards customers wanted during the modern global recession, MGM Resorts experimented with a complete overhaul of its loyal program design. Previously, rewards were not targeted, so people often wound up with undesired offerings, such as an older woman handed a pair of tickets to an Ultimate Fighting Championship. After the redesign, MGM Resorts personalized rewards to reflect customer activity data, to cater more accurately to customers' preferences [15]. If one customer spends more money visiting night clubs and attending shows rather than gambling, the resort offers entertainment coupons, for example. This strategic adjustment relies on MGM's loyalty program partnerships with affiliated companies, and it requires continuous adjustments. Without such a dynamic approach, the loyalty program might suffer from insufficient usage and activity by users; inactivity rates may reach as high as 85% in travel industry, much higher than the recommended benchmark rate of 25–35% [16]. When designed and managed effectively though, loyalty programs provide a comprehensive range of customer data, which in turn enhance the effectiveness of loyalty programs in a virtuous cycle. Still though, customers must always perceive that the value they receive from rewards justifies the use of their data.

Managerial Insights from Public Data. Marketing researchers benefit from using customer data that provide both managerial and empirical insights; a significant proportion of marketing research is driven by the customer data available within and outside firms, at the customer or more aggregate levels. For example, a study of brand positions over time used public, online, user-generated content and online reviews [17]. By performing data crawling and collection of unstructured online customer reviews from a variety of brands and product categories, the authors of that study suggest an enhanced method for mining marketing-relevant meaning from online chatter, which constitutes another channel for accessing customer-generated data. The resulting insights allow businesses and governments to improve their product and service delivery, reduce waste, and implement more effective policy changes.

Customer Perspectives on Data Value

Both popular and marketing publications report on the value that companies gain or lose from their customer data management practices. It is more challenging, and less common, to put a number on the value that customers gain or lose from those same customer data management practices. Some attempts to understand customers' perceptions of value rely on experiments that implement different pricing strategies. For example, AT&T introduced a pricing plan that charged customers an additional $30 per month for opting out of its Internet Preference Program that tracks customer activities [18]. This option generated substantial controversy though, leading AT&T to halt the program quickly, with the recognition that "We got an enormous amount of criticism from privacy advocates when we rolled it out … an ad-supported Internet service …. Privacy advocates screamed about that" [19]. But Robert Quinn, Senior Vice President of AT&T, also adds that "As the privacy revolution evolves, I think people are going to want more control, and maybe that's the pricing model that's ultimately what consumers want"—though what he means by control is not totally clear. Other companies are trying out similar data-sharing pricing strategies but with a more customer-friendly approach.

Sources of Customer Value

Personalized Offerings. Customers can benefit from sharing their data, because they enjoy improved product features and personalized recommendations and promotions. Even when customers are bombarded with constant requests to share their personal data, in certain circumstances, they may feel less violated and even could perceive the value of sharing [20]. The key conditions that mitigate the negative effect of data vulnerability concerns include customer perceptions of high transparency and control in the firm's data management practices. This suppressing effect occurs by reducing feelings of violation and increasing levels of trust. When customers sense that they have high control over the uses of their data, which also are transparent, they likely trust the firm to deliver value derived from their shared data.

Convenience and Financial Incentives. Convenience, such as the prospect of saving time by sharing personal data, and financial incentives also can enhance the value of registering for online services [21]. Many firms offer financial incentives to new customers, such as price promotions or complimentary rewards. Customers also have some a priori expectations about the monetary value of their personal data.

However, most critical is that firms reciprocate sufficient value to customers for the value the firm receives so that the customer perceives the exchange is fair. When customers perceive they are being treated unfairly they feel they need to punish the firm even at a cost to themselves. The emotional response to being treated unfairly is very powerful and can seriously undermine relationships, where it has been termed *relationship acid* due to the strength and pervasiveness of its negative effect [22]. Whereas, providing appropriate value in exchange for providing data can lead to customer gratitude and a desire to reciprocate, which can generate incremental sales, positive word-of-mouth (WOM), as well as strengthen the relationship over time through cycles of reciprocation [23].

Role of Loyalty Programs from Customers' Perspective

Beyond rewards, convenience, and financial incentives, customers gain and appreciate psychological benefits from sharing data, especially through loyalty programs. Similar to the firm's perspective, loyalty programs still gather substantial and varied data from customers, but from their perspective, the benefits and value of the firm's use of those data must outweigh the potential data privacy concerns. Some common benefits that customers achieve by joining loyalty programs include member-exclusive rewards (e.g., discounts, points, personalized promotions); to earn and reap the benefits, the customers need to stay engaged with the program and accrue more points that they may redeem for rewards. However, such engagement cannot be forced, so loyalty program managers need to ensure that customers receive the expected (if not more) value from the program. Customers can easily be distracted by other, competitive programs with better reward values, lose interest if their need for the company's products and services diminishes, or simply forget about their membership. It is mainly in customers' best interest to participate in loyalty programs for products and services that they use often, to maintain a continuous cycle of rewards.

Prior research into loyalty programs offers some good examples of what customers perceive as valuable. For example, a loyalty program with multiple tiers can generate psychological perceptions of higher status and gratitude among customers, though bystanders who observe the target customers receiving preferential treatment may perceive greater unfairness, with risks to firm performance [24]. Another benefit is a stronger sense of community. Customers are more loyal to programs that are communal in nature, rather than those that only offer financial incentives, due to their sense of community [25]. The resulting communal benefits include a feeling of belongingness and influence, integration and fulfillment of needs, and shared emotional connections with the brand community. Loyal customers who have strong relationships with the brand also are more forgiving of mistakes, even including product failures [26].

Putting a Price Tag on Customer Data

As AT&T's attempts to implement differential pricing to opt out of data sharing reveal, it is not easy to put a price on customer data. In particular, there are few comprehensive guides to the mechanisms and factors that drive customer data valuations. Concerns and expenses related to privacy data management are growing faster than the expansion of knowledge about how to deal with customer privacy. Therefore, we consider some initial efforts to integrate and extend extant knowledge related to customer privacy by proposing a conceptual framework. We also conducted a conjoint study to determine the value that customers place on their personal data. Ultimately then, we seek to provide guideline for managers and outline important avenues for research.

Customer Data Value Framework

To understand customer privacy value, we offer the conceptual framework in Exhibit 7.3, with two key inputs: data usage (how a company

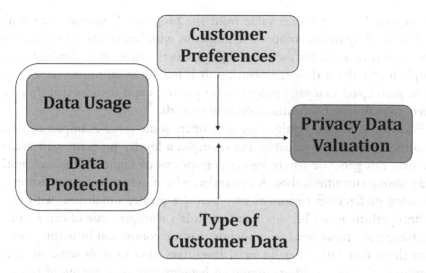

Exhibit 7.3   Customer privacy valuation framework

uses data acquired about its customers) and data protection practices (how a company provides data protection to its customers). These inputs are enhanced or suppressed by customer preferences, because customers express different desires and sensitivities for privacy protection, and by the type of data, to acknowledge that customers value some data types more than others. Finally, the output of the framework is privacy data valuation by a customer.

In line with this framework, firms must make two strategic decisions: how the data will be used and how they will be protected. Similarly, previous studies explore different types of privacy protections [27] and examine different levels of data usage, such as improving products or services, facilitating target marketing, and selling data to third parties. Some key mechanisms identified in extant literature include data vulnerability, a sense of violation, transparency, fairness, and trust [20, 28, 29]. The firm strategies can affect these mechanisms by determining how the privacy practice will implemented, such as by considering the differential needs of various customer segments in terms of their privacy preferences and which types of data are involved [25, 27]. Individual differences pertaining to privacy concerns, as well as previous experiences with data breaches also should be noted, together with demographic information. All these factors can influence customer valuations of their own data.

Analyzing Customer Valuations of Data

We use conjoint analysis to analyze the trade-offs among various features. This type of trade-off analysis is necessary, because a simple question that asks customers about the value they place on their privacy will invariably prompt a response of "high" or "invaluable!" Similarly, if a straightforward question asks about the extent to which they want to share their data, they likely will respond by saying "not at all." These extreme responses provide little insight, but a conjoint analysis can capture psychological ambivalence among opposing factors. Therefore, we conducted an online experiment with a conjoint design and analyzed individual preferences and values with a hierarchical Bayes estimation. (See Appendix for details of this study.)

Data Use Insights. All else being equal, consumers should prefer lower levels of data use by the firm, higher levels of data protection, and greater monetary payments for any data used. On an aggregate level, we estimate each data use valuation level by comparing it against a baseline level, at which the firm offers to delete personal data after every transaction. The incremental dollar value then indicates the amount that the consumer requires as compensation for the firm's continued use of his or her data. In Exhibit 7.4, we note that customers require $28 for the firm to use their data to inform their product improvements, rather than deleting their data after use. This payment would need to increase to $108 above the base case if the firm wanted to use their data for targeted promotions (e.g., targeted advertising), and then to $382 to allow the firm to sell their data to a third party.

Although improved efficiency, such as when they do not need to reenter their data to make a purchase, may have some value for customers, but customers appear to perceive it as a more meaningful benefit for the company, so they demand $28 in compensation. Moreover, communicating the intangible benefit of convenience achieved through data sharing may be insufficient, unless the firm can identify a group that particularly values convenience, more than other consumers. Learning from AT&T's failure and the results of our study, we suggest that instead of framing the opt-out choice as an additional charge, the firm should frame opting in as a financial reward, which may evoke greater acceptance. This prediction is consistent with

Analysis predicts that customers will require $28 for the firm to use their data to inform their product improvements, rather than deleting their data after use. This payment would need to increase to $108 above the base case if the firm wanted to use their data for targeted promotions (e.g., targeted advertising), and then to $382 to allow the firm to sell their data to a third party.

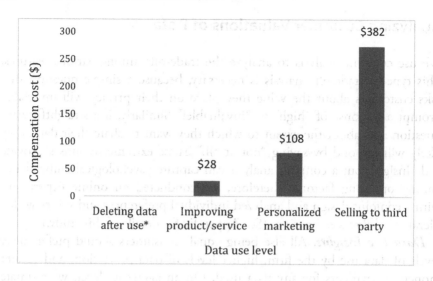

Exhibit 7.4 Data valuation for each data use level (*Note* The base case is for the firm to delete all personal customer data after completing the business transaction. The other situations are how much compensation an average customer requires to agree to a firm's specific data use relative to the base case)

prospect theory, which explains why losses loom larger than gains in people's minds [30].

Data Protection Insights. For the data protection levels, we evaluated in the study, utility dramatically increases when we move from no protection to suspicious activity monitoring, with a valuation of $326. The difference between no protection and a high level of protection with $1 million dollars of insurance is $411 (Exhibit 7.5). That is, if consumer data are protected by activity monitoring, it is worth $326 to customers, whereas a high level of insurance is worth an additional $85 versus monitoring. Financial data that reveal personal information may be worth additional insurance. Overall, protection seems to offer significant value to customers and can be a viable strategy for offsetting the value a customer wants to receive for a firm to use their data. To implement such a system, managers must determine specifically how much their customers value different data use levels and protection levels, then establish and target segments according to their privacy needs.

Data Type Insights. Data valuation also likely varies depending on how sensitive or critical the data are to customers. In our conjoint study, we also examine differential utilities of various data types, such as those related to retail shopping, financial services, and health care. The data type represents an additional factor, included in the conjoint design, together with data

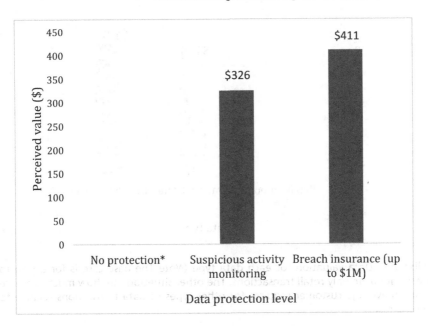

Exhibit 7.5 Valuation of data protection levels (*Note* The base case is for the firm to provide no data protection. The other situations are how much value an average customer perceives from receiving these extra protection services relative to the base case)

use, protection, and financial reward levels. Retail shopping data are relatively the least valued by customers, so they provide the reference point for calculating the relative utilities of the different data types. We find that the demanded compensation for financial data is $118 higher than that for retail shopping data; for healthcare data, it rises to $132 (Exhibit 7.6). In our sample financial and healthcare data had a relatively similar lift over general retail data, but it could be expected that this may vary dramatically across different customer groups (e.g., individuals with social stigmatized diseases or very high net worth individuals).

Privacy Needs-Based Segmentation. Because all customers are different, a latent class analysis can help explore which segments exist and how they vary in their preferences and valuations of privacy needs [31]. We identify four groups that express varying levels of attribute importance: convenience-seeking (24% of sample), data-conscious (13%), value-driven (27%), and privacy-protective (36%) segments. Each of these segments in turn can be positioned according to its levels of data use and data protection desires (Exhibit 7.7); they also show some notable demographic differences.

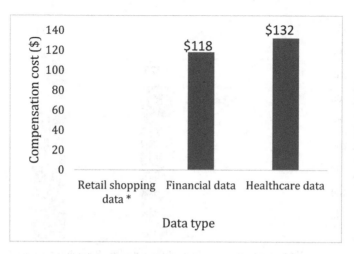

Exhibit 7.6 Data valuation for each data type (*Note* The base case is for a firm to collect data from only retail transactions. The other situations are how much compensation an average customer requires for other types of data transactions relative to the base case)

For example, the value-driven segment is youngest on average (36 years), and the data-conscious group consists of highest average age (43 years). The privacy-protective group includes the highest proportion of women (59%); the value-driven group contains the most men (64%) and features the lowest average income ($25,000–$49,000), whereas the convenience-seeking group has the highest average income ($50,000–$75,000).

Strategies for Sharing Data Value with Customers

The field of data privacy is changing rapidly and will lead to many iterations before the ultimate best practices emerge. Evidence from prior academic literature, our conjoint study, and real-world examples from the business press offer a consistent result: It is insufficient for firms to try to justify using customers' personal data simply by claiming that they will provide customers with added convenience and product improvements. This caution is especially relevant when the firm is monetizing customer data to a large degree, as Google and Facebook currently do. Time will tell if the consumer trends and the regulatory and financial impacts of GDPR on companies such as Google and Facebook include significant losses of revenue earned from targeted advertising.

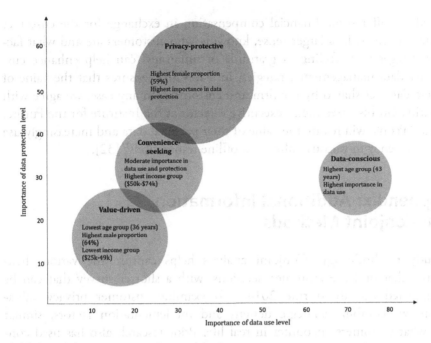

Exhibit 7.7 Privacy segmentation positioning map

At a minimum, managers must account for how their customers will respond to different data use levels, including whether they are collecting insights just to improve their offerings or if they plan to sell those data to third parties. Depending on the data types, people's sensitivity to the level of data use may vary greatly (e.g., general retail versus personal healthcare data). Customer data management efforts also should be accompanied by relevant, appropriate data protection programs to offset the data use costs, signal the firm's sincere interest in protecting consumers' data, suggest its ability to prevent or minimize the risk of data breaches, and reduce customers' perceptions of data vulnerability. These evaluations must include all the firm's specific customer segments, which exhibit substantial variance across industries, data types, and customers. Thus, privacy needs vary across different customer segments. In response, firms should position their customer segments according to their data use and data protection valuations. Assigning customers to these segments can help managers fine-tune their data management strategies. If a firm focuses mainly on convenience-seeking customers, then emphasizing product improvement efforts and personalized offerings could enable it to implement differential pricing strategies. If the firm's customers are mainly in the data-conscious group though, it likely needs to offer some financial compensation in exchange for the customer data it collects. In a larger sense, knowing who customers are and what factors trigger their feeling of gratitude or unfairness can help enhance customer data management strategies, in a way that ensures that the value of these data get shared by the firm and customers. In any case, we agree with Marina Gorbis' assessment, executive director at the Institute for the Future, that, "People will realize the value of their personal data and increasingly use it as currency in various online and offline transactions" [32].

Appendix: Additional Information on Conjoint Methods

Study methodology. Conjoint analysis helps capture the various heuristics that underlie customer decisions, with a shorter survey that can be completed quickly (Orme 2014). To examine customer privacy valuation, we combine strategic designs and implementation factors, similar to what customers encounter in real life. Prior research also has used conjoint approaches to observe customer data privacy valuations (Hann et al. 2007; Morey et al. 2015), but we extend this method by applying the latest configurations, including latent class analysis and hierarchical Bayes, which provides more fine-tuned insights into potential segments and utilities. Therefore, a choice-based conjoint study is well suited for our study purposes.

The online survey, using a conjoint design, allowed us to analyze individual preferences and values with a hierarchical Bayes estimation. With a random sample of the U.S. population (51% women), we obtained a mix of respondents who had or had not previously experienced a data breach. Respondents were instructed to choose the best alternative out of a set of concepts, composed of various combinations of the attribute levels for data type, use, protection, and valuation.

Choice-based conjoint analysis. Within the conjoint technique, multiple methods are available, including ratings-based, adaptive, and choice-based conjoint (CBC). Ratings-based conjoint, first introduced in the early 1970s, requires an exponentially increasing number of tasks by respondents as the number of attributes increases (Orme 2014). Thus experts raised concerns about respondent fatigue and decreased reliability with a ratings-based approach (Green and Srinivasan 1990). However, it can still yield useful results if the number of attributes is not too large. Adaptive conjoint was introduced instead helps respondents stay engaged by computationally adapting to each respondent's previous response and covering more attributes. Because adaptive conjoint is designed to examine main effects, not interactions, studies associated with pricing avoid it (Orme 2014). Finally, the most widely used method is CBC, which closely mimics customers' decision processes. Respondents review a set of product concepts that reflect a random combination of attribute levels; instead of ratings or rankings, they offer a choice. The series of choices, analyzed through multinomial logit analysis, reveal the partworth utilities of different attribute levels. Although this method is efficient and supports a larger number of attributes, it cannot reveal how strongly a respondent prefers one concept over another. Previously CBC entailed aggregate levels, but adding latent class and hierarchical Bayes estimation methods has meant that CBC now supports individual-level analyses. Therefore, "the majority of recent academic support and practical research applications are in favor of choice-based approaches." For our research purpose, we use Sawtooth software, a widely used conjoint tool in practice and research that accommodates such additional features. For additional technical details, please refer to the list of technical papers offered by Sawtooth (Orme 2000; Orme 2004).

References

1. Columbus, L. (2016, August 21). Roundup of Analytics, Big Data & BI Forecasts and Market Estimates, 2016. *Forbes Magazine.* www.forbes.com/sites/louiscolumbus/2016/08/20/roundup-of-analytics-big-data-bi-forecasts-and-market-estimates-2016/#27dfd8356f21.
2. Rainie, L., & Anderson, J. (2014, December 18). The Future of Privacy. *Pew Research Center: Internet, Science & Tech, Pew Research Center.* www.pewinternet.org/2014/12/18/future-of-privacy/.
3. Hsu, T. (2018, March 21). Users Abandon Facebook After Cambridge Analytica Findings. *The New York Times.* www.nytimes.com/2018/03/21/technology/users-abandon-facebook.html.
4. Meyersohn, N. (2018, March 22). Facebook's Stock Drops After Zuckerberg Apologizes. CNNMoney, Cable News Network. money.cnn.com/2018/03/22/news/companies/facebook-stock/index.html.
5. Gartner. (2017). *Gartner Forecasts Worldwide Security Spending Will Reach $96 Billion in 2018, Up 8 Percent from 2017.* Available at https://www.gartner.com/newsroom/id/3836563. Accessed June 27, 2018.
6. Wells, G. (2018, April 10). Facebook's Mark Zuckerberg Hints at Possibility of Paid Service. *The Wall Street Journal.* www.wsj.com/articles/facebooks-mark-zuckerberg-hints-at-possibility-of-paid-service-1523399467.

7. Madden, M. (2014, November 12). Public Perceptions of Privacy and Security in the Post-Snowden Era. *Pew Research Center: Internet, Science & Tech.* www.pewinternet.org/2014/11/12/public-privacy-perceptions/.

8. Mangalindan, J. P. (2012, July). Amazon's Recommendation Secret. *Fortune.* fortune.com/2012/07/30/amazons-recommendation-secret/.

9. Huet, E. (2015, June 17). How Airbnb Uses Big Data and Machine Learning to Guide Hosts to the Perfect Price. *Forbes Magazine.* www.forbes.com/sites/ellenhuet/2015/06/05/how-airbnb-uses-big-data-and-machine-learning-to-guide-hosts-to-the-perfect-price/#56126fb16d49.

10. Liberto, D. (2018, March 20). Facebook, Google Digital Ad Market Share Drops as Amazon Climbs. *Investopedia.* www.investopedia.com/news/facebook-google-digital-ad-market-share-drops-amazon-climbs/.

11. Brown, B., et al. (2017, March). Capturing Value from Your Customer Data. *McKinsey & Company.* www.mckinsey.com/business-functions/mckinsey-analytics/our-insights/capturing-value-from-your-customer-data.

12. Lewis, A., & McKone, D. (2016, October 21). To Get More Value from Your Data, Sell It. *Harvard Business Review.* hbr.org/2016/10/to-get-more-value-from-your-data-sell-it.

13. Reinsel, D., et al. (2017, March). Total WW Data to Reach 163ZB by 2025. *StorageNewsletter.* www.storagenewsletter.com/2017/04/05/total-ww-data-to-reach-163-zettabytes-by-2025-idc/.

14. Wood, A. (2016, March 15). Big Data: The Secret Weapon Behind Loyalty Programs. *Chief Marketer.* www.chiefmarketer.com/big-data-secret-weapon-behind-loyalty-programs/.

15. Benston, L. (2011, January). MGM Loyalty Program Will Reward Customers with Perks Not Imagined Previously. LasVegasSun.com. lasvegassun.com/news/2011/jan/06/you-could-pick-music/.

16. Meyer, J. (2017, November). Solving the Right Problems: How Retailers Can Leverage Data for Maximum Impact. *COLLOQUY,* 1–66.

17. Tirunillai, S., & Tellis, G. J. (2014). Mining Marketing Meaning from Online Chatter: Strategic Brand Analysis of Big Data Using Latent Dirichlet Allocation. *Journal of Marketing Research, 51*(4), 463–479.

18. Pressman, A. (2016, September 30). Here's Why AT&T Internet Customers Won't Pay Extra for Privacy Anymore. *Fortune.* fortune.com/2016/09/30/att-internet-fees-privacy/.

19. Smith, C. (2017). AT&T Could Start Charging You to Keep Your Private Data Private. Nypost.com. Available at https://nypost.com/2017/06/27/att-could-start-charging-you-to-keep-your-private-data-private/. Accessed June 27, 2018.

20. Martin, K. D., Borah, A., & Palmatier, R. W. (2017). Data Privacy: Effects on Customer and Firm Performance. *Journal of Marketing, 81*(1), 36–58.

21. Hann, I.-H., et al. (2007). Overcoming Online Information Privacy Concerns: An Information-Processing Theory Approach. *Journal of Management Information Systems, 24*(2), 13–42.

22. Samaha, S., Palmatier, R. W., & Dant, R. P. (2011, May). Poisoning Relationships: Perceive Unfairness in Channels of Distribution. *Journal of Marketing, 75*, 99–117.

23. Palmatier, R. W., Jarvis, C., Bechkoff, J., & Kardes, F. R. (2009, September). The Role of Customer Gratitude in Relationship Marketing. *Journal of Marketing, 73*, 1–18.

24. Steinhoff, L., & Palmatier, R. W. (2016). Understanding Loyalty Program Effectiveness: Managing Target and Bystander Effects. *Journal of the Academy of Marketing Science, 44*(1), 88–107.

25. Rosenbaum, M. S., Ostrom, A. L., & Kuntze, R. (2005). Loyalty Programs and a Sense of Community. *Journal of Services Marketing, 19*(4), 222–233.

26. McAlexander, J. H., Schouten, J. W., & Koenig, H. F. (2002). Building Brand Community. *Journal of Marketing, 66*(1), 38–54.

27. Hann, Il-Horn, et al. (2007). Overcoming Online Information Privacy Concerns: An Information-Processing Theory Approach. *Journal of Management Information Systems, 24*(2): 13–42.

28. Culnan, M. J., & Armstrong, P. K. (1999). Information Privacy Concerns, Procedural Fairness, and Impersonal Trust: An Empirical Investigation. *Organization Science, 10*(1), 104–115.

29. Morey, T., Forbath, T., & Schoop, A. (2015). Customer Data: Designing for Transparency and Trust. *Harvard Business Review, 93*(5), 96–105.

30. Kahneman, D., & Tversky, A. (2013). Prospect Theory: An Analysis of Decision Under Risk. In *Handbook of the Fundamentals of Financial Decision Making: Part I* (pp. 99–127).

31. Palmatier, R. W., & Sridhar, S. (2017). *Marketing Strategy: Based on First Principles and Data Analytics*. London: Palgrave.

32. Gorbis, M. (2014, December 18). *Pew Future of Privacy Report*, 57 www.pewinternet.org/2014/12/18/future-of-privacy/.

8

Data Privacy Marketing Audits, Benchmarking, and Metrics

Introduction

As the previous chapters make clear, privacy issues and strategies cannot be left to legal and IT departments. They are critical issues for marketing managers. A firm's policies and actions regarding the collection and use of big data (e.g., sales to third parties), customer targeting, advertising, and privacy policies all have significant impacts on customers' attitudes and behaviors, as well as vast financial implications. Even if a firm's legal department works to ensure that it is in compliance with privacy laws and regulations (Chapter 3), in-house lawyers often ignore the effects of the resulting policies on customers' perceptions—or at least they weight the potential legal consequences much more heavily than they do any customer and marketing effects.

Two years after the data breach at Facebook, Facebook's lawyers wrote to Christopher Wylie (the whistleblower in the Cambridge Analytica scandal) and "told him the data had been illicitly obtained and that ...it must be deleted immediately" [1]. Facebook's lawyers made no real effort to get the data back, beyond that one letter. Once the data breach became public, Facebook's stock price dropped, and a "Delete Facebook" social media trend grew. It thus appeared that legal coverage was the company's primary concern, not protecting customer privacy [1].

This chapter was researched with assistance from Sydney M. Zeldes, a marketing student at University of Washington.

© The Author(s) 2019
R. W. Palmatier and K. D. Martin, *The Intelligent Marketer's Guide to Data Privacy*,
https://doi.org/10.1007/978-3-030-03724-6_8

While legal departments focus on regulations and compliance, IT departments (appropriately) devote their attention to improving data security, preventing data breaches, and storing and managing large in-house databases. But this focus leaves little room for them to consider the role of data and privacy in day-to-day customer relationships. Existing IT audits are highly technical and inaccessible to managers who lack specialized computer science training, so they often have limited value in terms of informing the firm's overall business or marketing strategy.

Thus, the task cannot be left to other departments. Rather, today's firms need a marketing-focused, or better yet, customer-focused privacy strategy, and to help them develop one, this chapter outlines a process for conducting a customer-oriented data privacy audit, benchmarking exercise, which considers both firms' and customers' perspectives. We also identify the relevant metrics to include in a data privacy report card. To establish an effective marketing privacy strategy, every firm must understand its own privacy strengths and areas for improvement; the audit provides a structured process for collecting such information. It should proceed in parallel with legal and IT audits, such that the three types of audits achieve varying levels of detail, technical sophistication, and purposes, to serve their different audiences. A marketing privacy audit must be customer-centric; taking the customer's perspective is key to understanding how the firm's data collection, use, and privacy policies define its customer relationships. Such an audit also should carefully benchmark the firm's practices against industry standards and best practices, which results in a privacy scorecard that encapsulates a firm's position in the marketplace, key objectives, strategies, and actions.

Data Privacy Marketing Audit

Broadly, a data privacy marketing audit should comprise six key steps, as detailed in Exhibit 8.1, and results in a privacy scorecard. First, benchmarking efforts provide insights into industry standards, norms, and best practices. Moving beyond these analyses, it also is a good idea to look past proximal, same-industry exemplars to find outside-industry best practices that can be integrated into current practices. Second, a marketing-based privacy audit must adopt the customer perspective to determine how customers regard the firm's actions and policies. To capture the customer perspective, a number of tools are especially effective: qualitative interviews and focus groups, analyses of customer complaint data, lost customer analyses, social media screening, and online customer surveys. Third, analyses of the

Exhibit 8.1 A privacy scorecard resulting from data privacy marketing audit

---

1. Benchmark privacy marketing practices
    a. Competitors
    b. Industry best practices
    c. Outside industry exemplars
2. Customer perspective
    a. Competitors
    b. Customer complaints and lost customer analysis
    c. Customer survey
3. Regulatory/legal perspective
4. Employee perspective
    a. Qualitative interviews
    b. Employee survey
5. Data privacy marketing with SWOT analysis
    a. Strengths
    b. Weaknesses
    c. Opportunities
    d. Threats
6. Data privacy marketing: objectives, strategies, and action plan

---

external regulatory and legal environment can capture threats and requirements. Fourth, for an internal view, privacy audits need to solicit the opinions of the firm's employees, using qualitative interviews, focus groups, internal documents, or confidential employee surveys. Fifth, the combination of these data provide necessary insights to conduct a strengths, weaknesses, opportunities, and threats (SWOT) analysis of the firm's privacy practices across key stakeholders. Sixth, this analysis in turn provides a basis for developing the privacy marketing objectives and develop applicable defensive and offensive marketing-based data-privacy strategies, all of which must be consistent with the firm's overall mission, values, and positioning. After gathering and summarizing all these inputs, firms can develop a detailed action plan that suggests ways for it to achieve its strategic objectives pertaining to the privacy domain. Furthermore, the firm should establish metrics to track their progress toward these privacy marketing objectives.

Benchmarking Privacy Marketing Practices

Benchmarking efforts aim to develop a clear picture of a firm's practices, relative to those undertaken by competitors, industry leaders, and other exemplars. With a multidimensional evaluation, firms can capture relevant practices pertaining to data collection, use, and storage, as well as privacy policies, customer transparency and control, breach notifications, and compliance with regulations. That is, the firm's position, relative to other actors', should be defined according to each practice, to identify any potential gaps. The process might be the responsibility of a multifunctional team of internal employees or an external consultant; the goal is to limit the potential for bias in the results. In many cases, data privacy marketing audits can be performed in conjunction with (though still separate from) legal and IT audits, so that the firm can take advantage of any overlap in efforts and address the findings simultaneously.

Customer Perspectives

As we have reiterated in this book, the customer's perspective is critical, and it is especially so for data privacy marketing audits. Customers assess the firm's previous actions and policies, and when they appear purposefully ambiguous, such as relying on difficult-to-understand legal jargon or issuing overly vague privacy policies, their assessments will be unfavorable. To establish trust and long-term commitment among customers, transparency is critical, and without these elements, data breaches, disclosures, and lawsuits lead to irreparable damage to the firm's brand—in addition to the immediate financial effects [2].

Gaining insights into customers' perspective requires various techniques. For example, qualitative interviews or focus groups with customers (both current and potential) can give the firm a wide view and help it identify previously unexpressed concerns or less obvious secondary issues that managers might have missed. With this foundation, the firm can analyze its internal customer data, including complaints, lost customer analyses, and social media screening results, to identify the root causes for their concerns, including their suspicions regarding big data and privacy policies. These internal sources provide concentrated information about existing customers and their reactions (e.g., complaining, churning). The data are backward-looking, often arriving too late for the firm to act on the insights to retain specific customers, but they also provide likely hints of broader problems that many other customers might have faced, without ever complaining. Research suggests that this hidden group of "non-complaining" customers can be 26 times larger than the group of customers that voice their complaints [3].

Once the scope of issues and potential issues have been specified, the firm can develop an online survey to gauge people's perceptions of the focal firm, its competitors, and best practice exemplars, thus achieving a customer-centric perspective on the firm's performance across relevant dimensions. This survey should be anonymous, to limit response bias. Asking customers explicitly about a clearly identified firm can provoke them to provide misleadingly positive responses, because they seek to confirm their existing purchase decisions or want to be "nice" to the surveying firm, with which they have an ongoing relationship. The survey also should be multidimensional, spanning all potential issues identified by the qualitative interviews and analyses of internal data. In particular, surveys should include items to measure customers' perceptions of the firm's policies, the value they gain from its data management efforts, the fairness of its practices, and their data vulnerabilities due to the firm's actions. Regardless of what the firm is doing, or thinks it is doing, to protect customer data, customers' sense of the activities is what matters, and this perception often differs significantly from internal managers' views. Survey data should be generated from a random sample of respondents, using measurable scales, so the results can be generalized easily to an overall population and compared across different firms and over time. Key insights also might not stem from a result related to an actual level or score in one area but rather from how the customer's views change over time. Accurate trend information enables the firm to target its actions, aiming to be where the customer will be in a few years. However, surveys also have an inherent weakness; respondents usually are not "typical" customers, and it is questionable whether we can draw accurate causal inferences from cross-sectional, correlational survey data.

Regulatory and Legal Perspectives

Legal and regulatory environments constantly undergo change, often rapidly, when legislators and regulators attempt to demonstrate their reactivity to widely publicized current events. Furthermore, the complex legal environment is hard to predict and balance across global, country, and state levels (Chapter 3). Accordingly, audits of the regulatory and legal environment should leverage the specialized knowledge of the firm's legal department, which requires internal coordination. The legal department may have unique insights into how a regulatory change threatens to disrupt the firm's business model or revenue stream. For example, the recently enacted European privacy legislation, the General Data Protection Regulation (GDPR), is widely affecting the practices and policies of companies that earn profits from selling customer data. As these firms shift to comply with the stricter privacy controls, including mandates that require customers have the ability to opt out of sharing data but still use the services, they also are losing a key source of revenue, namely, selling targeted advertising. Still, legal requirements typically represent the minimum requirement; firms should want to exceed these requirements to build strong customer relationships and enjoy the benefits of enhanced customer loyalty in the long term.

Instead though, Facebook's response to GDPR seems to indicate its determination to do the least possible to meet legal requirements while still maintaining its revenue stream. But if Facebook users choose to share less personal data, targeted advertising could be less effective. In the words of Facebook CFO Dave Wehner, "there is a potential to impact targeting for our advertisers. Obviously if they are less able to target effectively, they'll get a lower ROI on their advertising campaigns. They'll then bid differently into the auction" [4]. Advertisers that fail to achieve success with their targeted advertising campaigns might spend less. The problem is not limited to Facebook either; Google uses similar data collection methods and targeting, and it relies heavily on advertising revenue. In response to potential legislative concerns about its business model, Facebook increased its spending on lobbying 50-fold since 2009, such that it spent $11.5 million in 2017, but still ranked third behind Amazon and Alphabet [5].

Employee Perspectives

Similar to the process used to gain customers' perspectives, firms should implement several methods to develop a clear picture of their internal employees' perspectives. Starting with qualitative interviews and focus groups of employees can reveal emerging or unexpected issues and ensure their inclusion in further assessments. Such interviews and focus groups should span all the departments in the firm, including employees in the legal, IT, customer service, sales, marketing, and R&D units, as well as those in all the geographically diverse offices that the firm maintains. In particular, insights from customer-facing employees should be sought actively, because they often have a clear sense of which data management policies or behaviors are irritating or worrying customers.

Relevant internal documents provide further information, in the form of published privacy policies; data collection, data use, opt-out, and targeting procedures; and third-party agreements for data sharing or sales. In this step, the firm should also attempt to match internal insights regarding its actions or inactions with external customer perceptions. In so doing, the firm gains a clearer sense of the true drivers of customers' sense of mistrust, vulnerability, or violation, as well as their long-term ramifications. For example, an analysis of the firm's privacy policy can provide a summary of its data policies and reveal which details remain ambiguous. A comparison of the privacy policies published by Apple [6], Google [7], and Facebook [8], as summarized in Exhibit 8.2, reveals both similarities and differences [9].

Exhibit 8.2 Comparison of Apple's, Google's, and Facebook's data privacy policies

	Apple [9]	Google [9]	Facebook [9]
Collection of personal data	"We also use personal information to help us create, develop, operate, deliver, and improve our products, services, content and advertising, and for loss prevention and anti-fraud purposes"	"Google collects information from use of its services such as: device information and location information. It also uses information customers give it when signing up for services"	"Facebook collects information on: things you do and information you provide, things others do and information they provide about you, your networks and connections, device information (including location and mobile number), information from linked websites, information from third party partners"
Collection of non-personal data	"We may collect information such as occupation, language, zip code, area code, unique device identifier, referrer URL, location, and the time zone where an Apple product is used so that we can better understand customer behavior and improve our products, services, and advertising"	"When showing you tailored ads, we will not associate an identifier from cookies or similar technologies with sensitive categories, such as those based on race, religion, sexual orientation or health"	"Non-personally identifiable information is used for advertising, measurement, and analytics services"
Cookies	"Apple's websites, online services, interactive applications, email messages, and advertisements may use 'cookies' and other technologies such as pixel tags and web beacons. These technologies help us better understand user behavior, tell us which parts of our websites people have visited, and facilitate and measure the effectiveness of advertisements and web searches"	"We use information collected from cookies and other technologies, like pixel tags, to improve your user experience and the overall quality of our services. One of the products we use to do this on our own services is Google Analytics"	Cookies used for: authentication, security, site, product integrity, advertising, recommendations, insights and measurement, site features, performance, analytics and research. Other parties use cookies in connection with Facebook products

(continued)

Exhibit 8.2 (continued)

	Apple [9]	Google [9]	Facebook [9]
Disclosure to third parties	"Personal information will only be shared by Apple to provide or improve our products, services and advertising; it will not be shared with third parties for their marketing purposes"	"We provide personal information to our affiliates or other trusted businesses or persons to process it for us"	"When you choose to use third-party apps, websites, or other services that use, or are integrated with, our Products, they can receive information about what you post or share"

Note Illustrative quotes came from each firm's 2018 online privacy policies

Apple uses personal information to develop and improve products; Google's privacy policy includes a similar indication that the company uses information collected from cookies to improve its services, including advertising, and it also allows third parties (including affiliates) to process data. Apple does not allow third parties to use data for their marketing purposes; Google and Facebook make no such claim. Facebook allows other parties to use cookies in connection with Facebook, and its privacy policy states that when users click on third-party apps integrated within Facebook, those third parties receive data about what the Facebook users post or share. Both Google and Facebook collect location information by default. Facebook collects information about users from third parties and linked websites. It also acknowledges that its data-sharing practices are for both its own and third parties' gains.

For some firms, confidential internal employee surveys give employees a channel to identify problems that may be in conflict with the firm's or senior managers' assertions and promises. Such surveys also should ask for recommendations for how to improve the firm's data management policies and customer relationships. These forms of employee input are very helpful, though employees also represent a prominent security risk, due to their lack of knowledge; one study suggests that only 12% of employees have sufficient knowledge to prevent a data breach [10]. Thus, the employee survey also can provide an overview of employee knowledge about data breach risk areas, including incident reporting, working remotely, access controls, data privacy policies, malware warning signs, social media, phishing, and cloud computing [11]. This internal audit of employee privacy knowledge can help the firm develop internal training programs and proactively prevent future data breaches (i.e., defensive strategies). Because internally focused data breach prevention efforts are often managed by the IT department, this portion of the survey should be well coordinated to prevent duplication of efforts.

Data Privacy Marketing with SWOT Analysis

The preceding data collection efforts provide marketers with the necessary insights to conduct a SWOT of the firm's privacy practices. This analysis aims to provide an unbiased picture of the firm's current position with regard to data management and privacy issues, the gaps between its existing position and the best practices, and a clear map of how the firm can improve and what it should aim to achieve, in anticipation of ongoing trends and likely changes. The structure of a SWOT analysis provides a helpful, clear basis for developing the firm's privacy marketing objectives, along with applicable defensive and offensive strategies. A good SWOT analysis establishes guidelines for allocating resources to reduce existing competitive gaps, avoid potential business threats, and leverage strengths to exploit business opportunities. However, the SWOT analysis must be customer-centric rather than firm-centric and acknowledge dramatic disruptions occurring in the focal domain.

> A critical requirement is that the SWOT analysis is customer-centric versus firm-centric and accounts for the dramatic disruption occurring in the privacy domain.

Consumer privacy preferences and new privacy legislation have great potential to disrupt existing business models. For example, GDPR will require substantial changes for the Internet Corporation for Assigned Names and Numbers (ICANN), which manages the Domain Name System (DNS), as well as the WHOIS public database, where people register their domain names, addresses, email addresses, and phone numbers. This system does not currently meet GDPR requirements; despite ICANN's request for more time to comply, it is likely its system will become obsolete. As security expert Brian Krebs explained, "WHOIS is probably the single most useful tool we have right now for tracking down cybercrooks and/or for disrupting their operations," but with the disruption caused by GDPR, it may not exist much longer [11].

Data Privacy Marketing: Objectives, Strategies, and Action Plans

After completing a SWOT analysis resulting in a privacy report card, managers can define their privacy objectives and develop effective defensive and offensive strategies; the landscape will be well defined. However, most firms treat data management and privacy issues as primarily defensive concerns, such that they focus on how to prevent data breaches, avoid customer complaints, or minimize bad press. When Citibank suffered a data breach, it hired thousands of IT personnel to prevent further breaches—likely a necessary response. But it made no notable changes to its customer-facing privacy policies. Thus, it continues to rank low in transparency and control scores, exposing it to substantial financial risk if it experiences another data breach and harms the customers whose concerns it has ignored [2]. As should be obvious, we believe that managers must move beyond such approaches to make the firm's data management and privacy policies a part of their offensive marketing strategy. Data practices can provide competitive advantages, evoking trust and loyalty among customers. This approach will be critical for firms that conduct most of their business online and accumulate vast amounts of data in the normal course of doing business.

For example, rather than burying an ambiguous privacy policy deep on a website, a competitive firm will make a concise, clear, accessible statement of its commitment to customers and their privacy. Costco Wholesale, which offers superlative transparency and control in its privacy practices, issues a strong commitment to customers in the first sentence of its privacy policy: "We respect your right to privacy. We do not sell, rent, share or disclose personal information to third parties without your prior consent.... In addition, your ability to make informed choices about the uses of your information is important to us." Throughout its clear, accessible, and easy-to-read policy, the company repeatedly affirms its commitment to respecting customers' privacy [12]. But for firms whose basic business model depends on earning revenue by selling customer data (e.g., Facebook and Google), other strategies might be required, including compensating customers for the use of their data or scrubbing those data carefully. Even if a firm does not make these changes willingly, regulators may force its hand, as exemplified by GDPR. Therefore, a detailed action plan should be regarded as a necessary output of the strategy to achieve objectives and implement strategies.

Furthermore, privacy audits should continue, with adequate frequency, reflecting the rate of change in regulations, customer expectations, and competitive actions. To monitor change, avoid time-consuming and costly audits, provide insight into progress in achieving objectives, and trigger new audits, managers need to identify and track key metrics that provide real-time insights.

Data Privacy Metrics

The final step, after completing a data privacy marketing audit, is to identify those metrics that can track the firm's progress toward its privacy objectives. These metrics might be added to an existing scorecard or used to develop a dedicated privacy scorecard, as a single source that senior managers can use to evaluate progress toward data privacy objectives. Capturing and reporting data systematically and over time has the significant advantage of providing a baseline norm (i.e., past month's score, competitor's score) for evaluating progress. Firm-specific metrics should reflect its SWOT analysis, objectives, and strategies, but other metrics will be relevant for nearly every firm. Ideally, the range of metrics will be diverse, spanning domains and times, external- and internal-facing measures, forward- and backward-looking assessments, and fast-acting versus more stable elements. We define some pertinent, appropriate metrics, along with their measurement methods, strengths, and weaknesses, as follows:

Privacy Policy Violations

- Incidents in which a firm violates its own privacy policy.
- Counts of violations, such as the number of times the firm failed to delete data it should have or tracked customers who opted out.
- Holds the firm to high standard, creates internal awareness of privacy practices and their weaknesses, pushes the firm to improve and address problems before they become public incidents.
- Customers do not necessarily see this measure, so even if it is important to the firm, it likely has no financial impact unless an incident is publicized.

Customer Data Incidents

- How many and what type of incidents related to customer data losses.
- Count and description of customer data incidents.
- Captures incidents of all sizes, can help prevent future incidents.
- Does not account for financial or reputation impacts.

Financial Impact of Data Incidents

- Total cost of discovery, response, and firm value loss after an incident.
- Stock price loss for public announcements, costs to fix incident, costs to prevent reoccurrence, cost of payouts to customers affected by data breach.
- Financial impact is an effective way to measure the size and magnitude of a breach, especially for public companies that experience a dip in stock price.
- Does not directly consider the customer perspective.

Customer Data Vulnerability

- Customer assessment of risks to their data. Drives customers' behaviors due to their concerns about data privacy.
- Measured with anonymous customer surveys. Can also use third-party audit results. Comparing results with competitive or industry leaders is helpful.
- Best assessment of actual data risks from customer's perceptive.
- Actual vulnerability is not necessarily the same as customers' perceived vulnerability.

Customer Trust

- Trust is the foundation of every great relationship and drives customers' behaviors due to their concerns for data privacy.
- Measured with anonymous customer surveys. Can also use third-party audit results. Comparing results with competitive or industry leaders is helpful.
- Indicates how successful the firm has been in differentiating itself with its privacy practices. If customers have a high level of trust, the firm has been successful in its offensive marketing strategies.
- Customer ratings of trust are subjective.

Customer Willingness to Share Data

- Percentage of customers who will share data with the company, given opt-out choices. Can also measure level of data customers will allow the company to access, with opt-out level options.
- Measured in anonymous customer surveys. Can also be measured by number/change in opt-ins versus opt-outs.
- Realistically captures customers willing to share data with the firm, with implications for customer trust and decisions to share data.
- Customers' stated willingness to share does not always align with their actual sharing behaviors. Customers may share data to get a benefit from the firm but might not be happy about it or trust the firm.

Customer Privacy Related Complaints

- Amount of customer complaints related to data and privacy concerns.
- Number of filed complaints with the firm's customer service center related to privacy.

- Complaints allow insight into privacy issues most salient with customers. Metric can also be associated with firm effectiveness in addressing these complaints.
- Only a small percentage of customers complain. This metric does not capture the customers who simply leave rather than complaining.

Lost Customers

- Number of customers who left the firm due to data privacy concerns.
- Lost customer analysis surveys or outgoing calls to previous customers, asking why they migrated.
- Lost customers are usually honest, and their feedback can help a firm improve.
- Knowing why customers left does not get the customers back. Data are backward looking, after customer has been lost.

Social Media Sentiment

- Tracking of social media sentiment to determine customer feelings and concerns.
- Text mining, sentiment analysis of social media text posts.
- Avoids survey biases and cost of surveys.
- Most posts or reviews on social media are from customers who are either very pleased or very displeased, so this metric does not capture the thoughts of customers who feel neutral or do not have strong opinions.

Time to Initiate Response

- Measures time to discover and respond to privacy incidents. Ideally, this metric continues decreasing as the firm fine-tunes its responses to data incidents and begins to respond to them more efficiently.
- Time from when incident occurred to discovery as well as to time taken to respond.
- If it takes a long time to respond, this metric will alert managers to investigate the cause of the delay. Fast responses can often suppress the negative impact of an event.
- Does not identify reason for delays.

Percentage of Staff Receiving Privacy Training

- Employees are one of the biggest privacy risks but can be mitigated by increasing the number of employees who receive privacy training.
- Percentage of employees in a company who have received training in privacy.
- The more employees are aware of data best practices and the threat of breaches, the more likely the firm will be to respond quickly and efficiently. The more staff have privacy training, the better.
- Need to make sure employees use the training they receive to protect data appropriately.

At the completion of a privacy audit, the metrics that are most relevant for that firm, according to its present position and planned offensive and defensive strategies, should be identified. Then the firm can institute an ongoing process for capturing and reporting privacy-related metrics, whether in a dedicated privacy dashboard or by integrating the metrics into the farm's existing reporting system.

In summary, a privacy marketing audit with accompanying metrics can be captured in a privacy scorecard, which is the aggregate document that captures a firm's position in the marketplace as well as key objectives, strategies, and plans to move the firm from where it is to where it wants to be regarding privacy issues. In addition, this scorecard should track the key metrics relevant for measuring progress for a firm's specific journey. Safeguards should be put in place to prevent bias in conducting both the audit as well as ongoing metrics to ensure managers' are seeing an accurate picture of their position and ongoing progress.

References

1. Cadwalladr, C. (2018). 'I Made Steve Bannon's Psychological Warfare Tool': Meet the Data War Whistleblower. *The Guardian*. Available at https://www.the-guardian.com/news/2018/mar/17/data-war-whistleblower-christopher-wylie-faceook-nix-bannon-trump. Accessed 22 June 2018.
2. Martin, K., Borah, A., & Palmatier, R. (2017). Research: A Strong Privacy Policy Can Save Your Company Millions. *Harvard Business Review*. Retrieved April 24, 2018, from https://hbr.org/2018/02/research-a-strong-privacy-policy-can-save-your-company-millions.
3. Markidan, L. (2015). Your Customers Aren't Telling You Everything. Here's What to Do About It. *Groove*. Available at https://www.groovehq.com/support/outbound-customer-service. Accessed June 22, 2018.

4. Wagner, K. (2018). Facebook Says Europe's New Data Privacy Rules Won't Hurt Its Business—Too Much. *Recode*. Available at https://www.recode.net/2018/4/25/17282612/facebook-gdpr-europe-user-growth-impact. Accessed June 22, 2018.

5. Bedoya, A. (2018). Opinion | Why Silicon Valley Lobbyists Love Big, Broad Privacy Bills. *Nytimes.com*. Available at https://www.nytimes.com/2018/04/11/opinion/silicon-valley-lobbyists-privacy.html. Accessed June 22, 2018.

6. Legal—Privacy Policy. (2018, May 22). Retrieved June 22, 2018, from https://www.apple.com/legal/privacy/en-ww/.

7. Privacy Policy—Privacy & Terms—Google. (2017, December 8). Retrieved June 22, 2018, from https://policies.google.com/privacy/archive/20171218.

8. Data Policy. (2018, April 19). Retrieved June 22, 2018, from https://www.facebook.com/policy.php.

9. Apple Legal. (2018). Legal—Privacy Policy—Apple. Available at https://www.apple.com/legal/privacy/. Accessed June 22, 2018; Facebook. (2018). Data Policy. Available at https://www.facebook.com/about/privacy/. Accessed June 22, 2018; Google. (2018). Privacy Policies. Available at https://policies.google.com/privacy#infocollect. Accessed June 22, 2018.

10. Ismail, N. (2016). Employees Lack Security Awareness—Information Age. *Information Age*. Retrieved April 22, 2018, from http://www.information-age.com/employees-security-awareness-levels-need-raised-123462944/.

11. Vaughan-Nichols, S. (2018). DNS Is About to Get into a World of Trouble with GDPR. *ZDNet*. Available at https://www.zdnet.com/article/dns-is-about-to-get-into-a-world-of-trouble-with-gdpr/. Accessed June 22, 2018.

12. Costco.com. (2018). Your Privacy Rights. *Costco*. Available at https://www.costco.com/privacy-policy.html. Accessed June 22, 2018.

9

Effective Privacy Marketing Strategies

Introduction

To help practicing business, managers develop marketing strategies that address emerging privacy issues; in this final chapter we synthesize the key points covered in previous chapters and organize them into six *Tenets for Effective Privacy Marketing*. These tenets encapsulate important, research-based insights; they also provide actionable recommendations that managers can use to address emerging privacy issues and to differentiate themselves from competitors while strengthening their customer relationships. Each tenet represents a key goal of an effective privacy strategy. By discussing the tenets in sequence, we highlight how they build on each other, reflecting a natural temporal ordering. Each tenet also is supported by multiple elements that are critical to its successful implementation.

For example, Tenet #1, *Avoid Customer Data Vulnerability*, reflects our recognition that customer data vulnerability is the primary element that undermines customer trust, leads to feelings of violation, and prompts negative customer behaviors (e.g., switching, reduced purchases, negative word of mouth [WOM]), so it should be the first priority in developing data privacy strategies (Chapters 1 and 2). Two key elements of reducing the effects of customer data vulnerability are (1) *data minimization*, or only capturing data that are needed and will be used, and (2) *empowerment through data transparency and control*, which suppresses the negative effects of customer vulnerability on customer behaviors (Chapters 2 and 5). In similar fashion, we offer six tenets

© The Author(s) 2019
R. W. Palmatier and K. D. Martin, *The Intelligent Marketer's Guide to Data Privacy*,
https://doi.org/10.1007/978-3-030-03724-6_9

Exhibit 9.1   Tenets for effective privacy marketing

Tenet #1	Avoid customer data vulnerability
	Data minimization
	Empowerment through transparency and control
Tenet #2	Balance big data and privacy
	Keep targeting from appearing creepy
	Build trust and strengthen relationships
Tenet #3	Share data value
	Fairness is key
	Guarding data helps
Tenet #4	Preempt privacy regulations
	Anticipate global consumer and legislative trends
	Develop simple and customer-centric policies
Tenet #5	Recover from data breaches
	Apology
	Protective compensation
Tenet #6	Audit for knowledge
	Create a privacy scorecard
	Identify and track privacy metrics

and their associated elements, as a sort of executive summary for developing and implementing an effective privacy marketing strategy. The six tenets and their constitutive elements are in Exhibit 9.1.

But these tenets also constitute more than just a summary of past research findings and the previous chapters. With them, we also attempt to extend prior insights by making a clear connection to specific business strategies and managerial actions. This linkage offers a critical bridge between academic empirical research and practicing managers' actions and decisions. That is, the tenets move beyond abstract theory and attempt to offer concrete guidance. Big data, artificial intelligence (AI) programs, data analytics, and other emerging technologies are challenging individual privacy to unimagined degrees and at a dramatically increasing pace, which is disrupting past business practices and moving firms into unmapped areas. We hope these tenets provide helpful guideposts for this journey.

Six Tenets for Effective Privacy Marketing

Tenet #1: Avoid Customer Data Vulnerability

The natural starting point for these guideposts is an unambiguous call to practitioners: Do not create or heighten customer vulnerability. Thus Tenet #1, *Avoid Customer Data Vulnerability*, resonates with the evidence that when companies collect, store, and use customers' personal information, it increases the potential for harm and thus people's feelings of vulnerability. As we described in Chapters 1 and 2, most negative customer effects resulting from data use derive from people's anxiety about the potential for damage or feelings of violation, rather than actual data misuses or financial and reputational harms. In our research, people report experiencing harm just from a company having access to their data—regardless of whether their personal information is analyzed, shared, or (mis)used. Throughout this book, we have argued that companies must take the strongest measures possible to avoid creating customer vulnerability. Thinking through how each element of the customer interaction raises, the potential for vulnerability is essential; it is not enough for marketers to focus only on actual or incurred consumer damages or harm.

Data Minimization. To avoid customer data vulnerability effectively, firms must only capture those data that are needed and will be used. In contrast, we see a widespread tendency of companies to gather as much detailed data from their customers as possible. Thus, companies that actively practice data minimization likely stand out, in positive ways. If they collect only the personal information that they need and intend to use to serve customers and personalize their experience better, companies engaged in data minimization also minimize the sense of vulnerability. In our in-depth discussion in Chapter 5, we offer concrete strategies that managers can use to inoculate their companies from data privacy harms. Notably, data minimization is a business practice, but it also can reflect the wider emphasis or firm culture that has been designed to put the customer's best interests first. By limiting their data collection efforts to only critical information, necessary to provide optimal customer experiences, companies ensure their own focus on strategies that truly offer value and enhance their marketing tactics and abilities. Such an approach also reduces costs and makes subsequent data analysis efforts more efficient.

Empowerment Through Data Transparency and Control. Beyond data minimization, managers can limit vulnerability by meaningfully engaging customers in data privacy practices. That is, by offering empowerment to customers, giving them transparency and control over their data, companies can suppress the negative effects of different types of vulnerability on subsequent customer behaviors. These are the practices we have promoted throughout this book, because they have proven effective across thousands of customer observations in our own research and also are consistent with vast research on gossip theory. Similar to data minimization practices, a dedication to embracing transparency and control is intuitive and actionable firm strategies. Yet these seemingly straightforward practices also have a great impact. When companies are transparent in their uses of personal data, they offer people some control over their personal data uses, the combination is potent. Customers feel empowered. Empowered customers trust a company more, feel less violated when their information is shared unintentionally, and report perceptions of greater fairness, loyalty, and desire to share favorable WOM. Beyond these focal firm impacts, as we describe in Chapter 6, transparency and control can shield a company from harm if a close rival experiences a data breach. Empowering customers through transparency and control, coupled with a company mindset to embrace data minimization, are proven methods to limit customer vulnerability, as promoted in Tenet #1.

Tenet #2: Balance Big Data and Privacy

With these critical building blocks in mind, the firm can balance the tensions that arise between leveraging big data's benefits and managing its inherent risks. The imperative to *Balance Big Data and Privacy* represents Tenet #2. As we elaborate in Chapter 4, customer data applications and analytics provide firms with powerful platforms they can use to enhance value creation, communication, and delivery processes. We encourage managers to harness the opportunities for business success embodied by these advances. Yet the precision with which companies can identify and target customers, based on their deep understanding of them, has the potential to appear too personal. As we highlight, research shows that the temptation to employ highly specific, precision targeting can come to seem creepy to customers, who get the sense that the firm is using big data to stalk them digitally.

Keep Targeting from Appearing Creepy. The first key element underpinning Tenet #2 thus calls on managers to avoid giving off a creepy vibe by implying the possession of intimate customer knowledge, beyond what could be gained readily through normal business interactions. The same technologies that identify and reward the company's best performing customers, to keep them loyal, also might eliminate, discourage, or create obstacles that encourage the company's worst performing customers to leave. Alienation is but one method firms might use to punish or exclude poorly performing customers. Many other unintended consequences of novel big data-enabled technologies also exist, in reality or in theory. When intelligent managers understand these practices and their potential for misuse, they can better prevent the emergence of unintended consequences. However, these tensions represent an ongoing balancing act for the firm. As Chapter 4 identifies, applications and technologies hold myriad benefits and risks for both customers and the company. Many emerging technologies (e.g., AI, machine learning) remove human emotional elements, such as empathy, which can have negative consequences on long-term relationships.

Build Trust and Strengthen Relationships. We recommend avoiding negative actions that result in consumer alienation or precision-targeted stalking; we recommend other company behaviors to cultivate and achieve the balancing act represented in Tenet #2. That is, managers must take positive, proactive measures to increase customers' trust in and relationships with the company. This second essential element of Tenet #2 involves a customer-centric mindset that underlies everything the company does—including but not limited to its big data-enabled technologies and practices. When customers and their best interests are focal and central in the big picture of the firm's strategy, spanning all elements of its value creation, communication, and delivery, the uses of technologies in ways that foster trust and strengthen customer–company bonds are likely to be natural outgrowths.

Consider Microsoft, a company that scored well with regard to its data privacy practices in our own research and that holds a top 10 ranking on *Forbes'* list of the World's Most Reputable Companies [1]. It is a rare example of a technology company that uses cutting-edge applications and analyzes simultaneously to embrace customer data privacy and engage with the social and global issues its customers care about. Microsoft's customer focus both builds trust and strengthens relationships. Similar to other "Most Reputable" peers, such as the Walt Disney Company and Sony, Microsoft has mastered the art of balancing world-class abilities to harness big data insights to create customer value, without ever wavering from its commitment to preserve customer privacy, which are the cornerstones of the balancing act described in Tenet #2.

Tenet #3: Share Data Value

Firms often gain significant benefits by collecting and using customer data, and failing to share a portion of that received value with customers is an unsustainable strategy. The idea that customers might gain value from better designed and targeted offers or advertisements is not sufficient, because it leads to customer perceptions of unfairness, which erodes the customer–company relationship through its destructive power (Chapter 7). When a company fails to share the immense value it receives from selling customer data to third parties, its actions appear even more egregious. Compensating customers by offering free services may be an effective short-term strategy but ultimately undermines customer relationships and is susceptible to competitive attacks from rivals that commit to and proactively promote the practice of not selling data.

A firm thus should evaluate how best to *Share Data Value* with customers (Tenet #3) after addressing the *Balance [of] Big Data and Privacy* (Tenet #2). Only after the firm has determined the breadth of its data collection and planned uses can it identify the value of those data and the customer's fair share. Companies already have many ways to share benefits with customers, such as free services, loyalty program rewards, and data protections, but in the future, we expect companies to offer more creative programs to share the value of these data. Imagine a "data profit sharing" arrangement for example, in which customers receive some predetermined portion of the profit that the firm earns from selling their data to third parties. Google or Facebook might send you a quarterly data royalty check for the value they have extracted from your data. Just as book authors often receive royalties from publishing companies, firms like Facebook and Google reasonably could pay a percentage of the profits they earn, in direct acknowledgment of the value of the contributed data. Such actions would transform the firm and its customers into partners, devoted to monetizing data. As an additional benefit, customers likely would cooperate and provide more *reliable, accurate,* and *valuable* data elements to a partner in a data monetization agreement. At the very least, they may be prone to opt in for more data uses. To advance to a point that such agreements might be possible and implementable, Tenet #3 also establishes two components that firms must undertake immediately.

Fairness Is Key. As an important element of Tenet #3, firms must actively seek to appear fair to customers. As an important stipulation however, fairness always must be evaluated from the customer's perspective. It will vary significantly across different customers and for different types of customer data. Unfairness has been shown to be "relationship acid," which motivates customers to punish firms even at a cost to themselves (Chapter 7). As competitors begin to offer customers compensation for their data, past data practices may begin to be perceived as unfair with accompanying customer punishing behaviors.

Guarding Data Helps. Clear protection of their data has significant value for many customers, because it reduces their data vulnerability and feelings of anxiety but also represents an offsetting investment by the firm for the value it has received. Guarding personal information (e.g., protection, monitoring, insurance) offers customers another benefit, because the company assumes some of the customer's risk of a potential data breach. Firms are the entities that collect, store, and provide data security, so data breaches typically are the company's fault. Yet the pain and inconvenience of monitoring or experiencing identity theft often solely affects the customer. A sound plan for guarding data sends a strong signal that the company has confidence in its handling of customer data and even is willing to share the risk and cost of a data breach.

Tenet #4: Preempt Privacy Regulations

Firms operate in novel, constantly changing environments with updated privacy regulations, and those shifts must be accounted for in any effective privacy marketing strategy, as outlined in Chapter 3. Firms might wait until the legislation is enacted, do the minimum that is legally required, fight in the courts, spend on lobbyists to thwart legislation, and treat global customers differently, based on distinct international requirements. In the short term, this approach may maximize firm profits. But it also sends a clear message that the firm is not customer-centric in its policies, undermines customer trust, and implies that the firm only reacts to legal mandates. In time, ongoing lawsuits and unfavorable press coverage might damage the firm's brand, perceived trustworthiness, and customer relationships.

Anticipate Global Legislative and Consumer Trends. Instead, Tenet #4 requires establishing, in advance, universal customer-centric policies that meet and typically exceed all legal mandates. Such an approach puts the firm ahead of competitors, exceeds people's expectations, and simultaneously generates customer goodwill (trust, relational bonds). These benefits combine to evoke long-term customer loyalty. Furthermore, such an approach prevents competitors from using their data privacy capabilities as a differential advantage. Applying global requirements (e.g., GDPR) to all customers not only is operationally easier to execute but also ensures fair treatment of customers across geographies. As Tenet #3 acknowledges, fairness is key, whereas unfairness can destroy customer relationships. Such long-term benefits outweigh the temporary benefits of short-term gains achieved from less forward-looking efforts.

Develop Simple and Customer-Centric Policies. Some firms are finding creative ways to meet GDPR mandates with complicated and clever, if not downright sneaky, notices and opt-out policies that their in-house legal counsels have determined meet the letter of the law but that fail to conform to the spirit of the law. Even if a company determines that its clever practices protect it from potentially large penalties (i.e., 4% of annual revenue in the case of GDPR), customers often are frustrated and see through this purposeful complexity. Simple, customer-centric policies enhance trust and allow customers to manage their data uses, easily and uniquely. The resulting feelings of control help suppress the negative effects of data vulnerabilities (Tenet #1).

Some firms and managers argue that customers do not really care about their privacy—or at least they behave so, as exhibited by the privacy paradox. So why should firms assume such stringent approaches to privacy (Chapter 1)? We argue that this view is risky. Consumer trends do not move in smooth, linear fashion; in some cases, people do not know what is important to them until they experience it in the market. For example, early evidence suggested there was limited customer interest in a portable music player or overnight letter delivery, but once consumers had an opportunity to listen to music through their Sony Walkman or use the services of FedEx, different performance features became expected, and demand grew quickly. Along similar lines, forward-looking search engines, phone companies, social media, and retailers might promise never to use or sell data unless a customer clearly granted permission, and in return, would provide those customers with a share of the data value. As *Forbes* recently predicted, consumers' concerns about privacy suggest "Will 2018 be the start of major privacy pushback? Yes" [2]. Overall, we anticipate that customer expectations and demands about data privacy, transparency, control, and value sharing will change dramatically over the next decade as firms increasingly highlight their data privacy policies as a performance attribute of their offerings, and as customer expectations evolve. Only time will tell, but we believe it is short-sighted *not* to anticipate dramatic changes as legislators and consumers catch up to the capabilities and concerns involved in new technologies.

Finally, another advantage of providing customers simple, easy systems for managing their individual data with regard to specific product/service offerings is that customers may help firms improve their targeting. For example, a customer of a BMW sedan may opt out of everything but maintenance information after purchasing a new car, then four years later could opt back into receiving marketing information, because this consumer is thinking of purchasing a new BMW in the next year, so this information is now valuable. A dynamic, customer-centric data management capability allows customers to manage their flow of information to match their needs and also grants them a feeling of control in their interactions with the firm.

Tenet #5: Recover from Data Breaches

Although we wrote this book to bestow managers with the knowledge, skills, and strategies needed to promote world-class data privacy practices in their organizations, the harsh reality is that many of the world's very best companies will suffer a data breach. Therefore, we wish we did not need to include it, but we must propose Tenet #5, the imperative to *Recover from Data Breaches,* to prepare firms for this eventuality. As we described in Chapter 6, our own research demonstrates that no modern organization truly is protected from a data breach. Combining these insights with global statistics on data breach costs and consequences [3] reveals that organizations and firms of all types must have a data breach recovery strategy in place. This strategy should specify how the firm will respond, beyond the parameters of IT or the security system specifications. A good data breach recovery plan should speak directly to how the firm will communicate with its customers, as well as how those customers will be compensated for any direct or potential losses incurred.

Apology. The first essential element of Tenet #5 involves an organizational apology. Offering customers an apology does not imply guilt or attribution for the breach. An apology is not an acknowledgment of blame. Rather, apologies are useful, positive ways to let customers know that the company recognizes their pain and cares about the potential harm that may result from any data breach event, regardless of the cause. Chapter 6 draws insights from the ways in which medical professionals convey bad news to patients suffering a difficult diagnosis. Because data breaches represent unknown harm (i.e., customers may never incur damage; they may incur damage far into the future; identity theft can be difficult to monitor), they often resemble an uncertain, serious medical situation more than a product failure or recall—crisis events that company leaders are better equipped to manage. Therefore, when managers adopt an empathetic data breach response, including a heartfelt apology, customers react less negatively and are less likely to defect or switch to a close rival. As the evidence in Chapter 6 reveals, an apology, empathy, or atonement can restore customer relationships. Thus, we reaffirm that a company apology following a data breach will be well received by customers, markets, and investors, with weaker negative effects in the long run.

Protective Compensation. Many companies began offering complementary credit monitoring solutions following a data breach, so this practice has grown to become widely expected by both customers and investors evaluating firm stock prices. In this case, protective compensation through credit monitoring services represents a viable option for firms attempting to restore the equity exchange and provide at least some reduction in vulnerability (Tenet #1). Likewise, because a data breach represents a unique type of customer harm, a protective compensation strategy could create some beneficial recovery outcomes. That is, because in the contemporary business landscape, companies and customers frequently are plagued by data breaches, free credit protection seemingly has become the expected minimum form of reparation. It signals the benevolence that customers seek, and it provides a form of vulnerability mitigation, even if most people understand the limits of these protective services.

Our research also strongly advocates for the creation and dissemination of data breach recovery plans so that managers can respond, quickly and effectively, to a breach should one occur. Company apologies and a tangible offering of protective compensation are two critical elements of such a plan. Even if we hope that organizations have no use for such strategies, the unfortunate reality is that breaches will continue to occur. Firms have the elements required to recover from a breach, as one specific data privacy event. But thinking with a big-picture, long-term mindset also means that firms should establish a guiding framework, to assess all marketing practices with an eye toward data privacy. For this reason, our sixth and culminating tenet embraces comprehensive firm data privacy evaluations, conducted through a unique auditing program that we outline.

Tenet #6: Audit for Knowledge

Tenet #6, *Audit for Knowledge*, highlights that developing an effective privacy marketing strategy demands that a firm understands where it stands in customers' and employees' minds. The privacy audit described in Chapter 8 outlines a process for capturing an unbiased picture of where a firm is positioned relative to competitors and best-in-class exemplars.

Create a Privacy Scorecard. This critical step moves beyond managers' platitudes or wishful thinking and provides an unvarnished picture of their firm's position, including its strengths, weaknesses, opportunities, and threats (SWOT). From this baseline snapshot of the firm's present position, as well as the guidance gleaned from applying the previous five tenets, a firm can develop key objectives and action plans, which will constitute its overall privacy strategy.

Identify and Track Privacy Metrics. Specific measures will be most relevant for each firm's unique strategy, as described in detail in Chapter 8. Ideally, these metrics also can be used to track the progress, from the firm's existing position to its desired position, and will include customer-oriented and employee-focused measures, as well at other relevant assessments of the firm's strategy. The importance of customer-oriented metrics is fairly obvious, but many firms fail to recognize the multiple benefits of employee-focused metrics, especially in the privacy domain. First, employees often have unique insights into the underlying causes of customer complaints (e.g., firm policies), as well as potential solutions. Second, capturing and using customer data, and the related challenges in the privacy domain in general, can have an impact on an employee's underlying values or even morals; when violated, it might even suppress employee motivation or lead to turnover. For example, Google employees pushed back on several initiatives, resulting in the firm's exit from a U.S. Defense Department project, though some employees still left the company in protest [4].

In general, the privacy audit and resulting privacy scorecard can catalyze the establishment of clear processes for using big data. The process, company privacy policies, and all privacy-related actions should constitute key elements of the firm's overall marketing strategy. Throughout this book, we have argued that privacy cannot be relegated to the company's IT or legal domains—both of which are very important but insufficient alone. Ultimately, data privacy decisions fundamentally affect customer trust, relationships, long-term loyalty, and financial performance. The related marketing strategies should be more than just defensive in nature; they should reflect and constitute key aspects of a firm's positioning strategy, offering opportunities for building competitive advantages.

The Future of Data Privacy Marketing

It is very difficult, if not impossible, to make predictions with any degree of confidence, but most knowledgeable experts agree that big data and the resulting need for privacy policies will not remain stagnant. As a few clear takeaways, we note that legislative and regulatory changes are nearly guaranteed. It is just a question of time and how extensive or draconian they will be. We anticipate a sort of global convergence toward the most stringent requirements across most major economic markets. Firms will want to access these large, wealthy markets, and it will be difficult to differentiate privacy standards across them.

Another prediction that we are fairly confident about is that consumers will issue louder, stronger demands for transparency, control, and compensation for their personal data. In other words, the privacy paradox will be resolved. Consumers will begin to behave in ways consistent with their stated preference not to share their data. Technological advances have enabled big data, precision targeting, and online marketing, and in many cases, they likely will solve the ensuing privacy issues too. What if, for example, every consumer had her or his own standardized privacy dashboard, providing control over all aspects of how firms track, collect, or use their personal data? Each person could manage every firm's interface with him or her, according to privacy preferences, opting out of any sales of data to third parties with the click of one opt-out button. How many consumers would make that selection? We predict that many of them would, especially if no acceptable compensation is being offered. In contrast, most firms currently make the process of opting out very complex, with confusing statements and difficult-to-find options. Each company uses a different process, making it very laborious for a consumer to manage her or his data usage. In turn, many people do not bother, leading to something like a privacy paradox. With the right systems in place though, the big data and privacy worlds could endure the modern radical transformations and effectively dissolve the privacy paradox.

It also is hard to predict the speed with which these changes will occur, and it must be acknowledged that there are many firms with deep pockets, extensive lobbying networks, and significant motivation to prevent or at least slow down this process. Still, we believe entrepreneurs and firms without a stake in the status quo ultimately will address this latent customer need. It may take some major, highly publicized event to sway elected officials, media, mainstream popular opinion, and even the negatively affected business leaders before all firms to become sufficiently motivated to take real action and grant enough control to consumers.

These ongoing changes make it hard to write a book on the subject; they make it especially hard for us to come to the end of it. Once we conclude and put the book into print, new laws, changes in consumer expectations, and emerging empirical research will offer additional insight on these topics almost immediately. In addition, treating big data and privacy policies as a focal domain for proactive marketing strategy is relatively new. We accordingly expect many new insights and processes to emerge. With this anticipated evolution in mind, we plan to continue posting new research on this subject for use by our readers at SAMSinstitute.com.[1] For example, we are working to launch a concise, well-supported methodology for generating privacy scorecards that will be available on this website in the near future.

---

[1] The SAMS Institute (Sales and Marketing Strategy Institute) is a voluntary association of professors and interested business executives focused on *linking academics to business for knowledge.*

References

1. Valet, V. (2018, March 15). The World's Most Reputable Companies 2018. *Forbes.* Available at https://www.forbes.com/sites/vickyvalet/2018/03/15/the-worlds-most-reputable-companies-2018/#55b1631126d5. Accessed September 18, 2018.
2. Andriole, S. (2018, January 3). *Big Trouble for Facebook, Amazon, Google and Apple in 2018.* https://www.forbes.com/sites/steveandriole/2018/01/03/big-trouble-for-facebook-amazon-google-apple-in-2018/#67e6d1794d87.
3. For specific statistics and dollar amounts, please see IBM Security, Ponemon Institute. (2017). *2017 Cost of Data Breach Study.* Available at https://www.ibm.com/security/data-breach.
4. Gray, S. (2018, May 14). *Report: Google Employees Resigning over Controversial Pentagon Contract.* Available at http://fortune.com/2018/05/14/report-google-employees-resign-pentagon-contract/.

Company Index

© The Editor(s) (if applicable) and The Author(s), under exclusive
licence to Springer Nature Switzerland AG 2019
R. W. Palmatier and K. D. Martin,
The Intelligent Marketer's Guide to Data Privacy,
https://doi.org/10.1007/978-3-030-03724-6

Subject Index

© The Editor(s) (if applicable) and The Author(s), under exclusive licence to Springer Nature Switzerland AG 2019
R. W. Palmatier and K. D. Martin,
The Intelligent Marketer's Guide to Data Privacy,
https://doi.org/10.1007/978-3-030-03724-6